**DRILL HALL LIBRARY
MEDWAY**

METHODS IN MOLECULAR BIOLOGY™

Series Editor
**John M. Walker
School of Life Sciences
University of Hertfordshire
Hatfield, Hertfordshire, AL10 9AB, UK**

For further volumes:
http://www.springer.com/series/7651

RNA Abundance Analysis

Methods and Protocols

Edited by

Hailing Jin

*Department of Plant Pathology & Microbiology, Institute for Integrative Genome Biology,
University of California, Riverside, CA, USA*

Walter Gassmann

*Division of Plant Sciences, Christopher S. Bond Life Sciences Center & Interdisciplinary Plant Group,
University of Missouri, Columbia, MO, USA*

Editors
Hailing Jin
Department of Plant Pathology & Microbiology
Institute for Integrative Genome Biology
University of California
Riverside, CA, USA

Walter Gassmann
Division of Plant Sciences
CS Bond Life Sciences Center
& Interdisciplinary Plant Group
University of Missouri
Columbia, MO, USA

ISSN 1064-3745
ISSN 1940-6029 (electronic)
ISBN 978-1-61779-838-2
ISBN 978-1-61779-839-9 (ebook)
DOI 10.1007/978-1-61779-839-9
Springer New York Heidelberg Dordrecht London

Library of Congress Control Number: 2012936643

© Springer Science+Business Media, LLC 2012
This work is subject to copyright. All rights are reserved by the Publisher, whether the whole or part of the material is concerned, specifically the rights of translation, reprinting, reuse of illustrations, recitation, broadcasting, reproduction on microfilms or in any other physical way, and transmission or information storage and retrieval, electronic adaptation, computer software, or by similar or dissimilar methodology now known or hereafter developed. Exempted from this legal reservation are brief excerpts in connection with reviews or scholarly analysis or material supplied specifically for the purpose of being entered and executed on a computer system, for exclusive use by the purchaser of the work. Duplication of this publication or parts thereof is permitted only under the provisions of the Copyright Law of the Publisher's location, in its current version, and permission for use must always be obtained from Springer. Permissions for use may be obtained through RightsLink at the Copyright Clearance Center. Violations are liable to prosecution under the respective Copyright Law.
The use of general descriptive names, registered names, trademarks, service marks, etc. in this publication does not imply, even in the absence of a specific statement, that such names are exempt from the relevant protective laws and regulations and therefore free for general use.
While the advice and information in this book are believed to be true and accurate at the date of publication, neither the authors nor the editors nor the publisher can accept any legal responsibility for any errors or omissions that may be made. The publisher makes no warranty, express or implied, with respect to the material contained herein.

Printed on acid-free paper

Humana Press is a brand of Springer
Springer is part of Springer Science+Business Media (www.springer.com)

Preface

RNA abundance analysis is one of the most important approaches for gene expression studies in the field of molecular biology. Rapid progress in modern technology has empowered us to examine RNA expression more accurately and efficiently. A technical review of the methodology of RNA abundance analysis is timely. This book covers a wide range of techniques on RNA extraction, detection, quantification, visualization, and genome-wide profiling, from conventional methods (e.g., Chaps. 1, 2, and 5) to state-of-the-art high-throughput approaches (e.g., Chaps. 3, 4, 12, and 14). We include detailed techniques to examine mRNAs, small noncoding RNAs, protein-associated small RNAs, sulfur-containing RNAs, viral and satellite RNAs, RNA isoforms, and alternatively spliced RNA variants from various organisms. The computational data processing for genome-wide datasets is also discussed (Chaps. 15–17). Collectively, these methods should provide helpful guidance to biologists in their gene expression and regulation studies.

The beginning of many RNA studies is the isolation of RNAs. We have included methods for extracting RNAs from cells and tissues of plants, fungi, insects, and parasites (Chaps. 3, 4, 6, and 12). This can be challenging if one wishes to address a process that is limited to a few cells within a tissue or organism, e.g., cancerous cells within healthy tissue (1) or host–pathogen interactions (2). Chapters 3 and 6 present the methods to extract RNAs from microscopic animal and plant tissues, respectively. In plants, additional problems include the uniformity of cell types to the naked eye and the presence of a cell wall, characteristics of which can interfere with downstream processing of cellular samples. Chapter 6 describes a method that combines high spatial resolution with the aid of laser capture microdissection technology with a proven protocol to isolate high-quality RNA from the embedded and sectioned tissue. This material can then be used for new gene discovery or genome-enabled gene expression studies.

If the gene of interest is already known and spatial expression patterns within a specific organ or tissue need to be established, in situ hybridization is a method of choice (e.g., (3)). Plants can present a unique challenge. Chapters 5 and 11 describe in situ hybridization methods to detect the expression of mRNAs and microRNAs at the cellular level, which deliver highly gene-specific expression data within the context of the organismal structure.

Just as our knowledge of the functional roles of RNA has expanded in recent years, it is also known that the chemistry of RNA is diverse. Cells naturally possess sulfur-containing tRNAs that incorporate thiouridines (4), and this chemical modification can also be used experimentally to probe structure–function relationships of catalytic or metal-binding RNAs (e.g., (5, 6)). In Chap. 8, a very efficient method that allows the isolation of in vivo or in vitro sulfur-modified RNA is described.

This volume contains several chapters that describe high-throughput genome-wide next-generation sequencing approaches to study RNA-related parameters in organisms. However, often a focused detection of individual transcripts of known sequence is needed. Chapter 9 describes a highly efficient, simple, and quantitative method using RNA protection assays to measure multiple RNA isoforms arising from a single transcriptional unit. One good example is viruses, which, with a very compact genome, need to encode multiple proteins (e.g., (7)). Viruses accomplish this feat by processing the same RNA multiple ways using alternative sites for transcription initiation, splicing, and poly-adenylation. The method described in Chap. 9 was developed mainly with parvoviruses but is applicable to many situations where a single gene gives rise to multiple mRNA isoforms.

Chapter 7 follows by focusing on the analysis of genome-wide alternative splicing using RNA-Seq. Because many alternative transcripts are not yet annotated even with such well-developed genome resources as those for the plant *Arabidopsis thaliana* (e.g., (8)), capturing the spectrum of alternative transcripts is still a complex challenge that this chapter addresses with a tested series of programs.

Chapter 1 tackles the slightly less complex question of which genes are actually expressed in a given sample and at what level. The method described in this chapter, SuperSAGE, is a further development of the powerful SAGE technology (9) that uses short tags to identify which mRNAs are present in a biological sample and quantifies expression levels by counting the number of tags for each gene. SuperSAGE combines longer tags for improved gene assignment with high-throughput sequencing (10).

Isolation and profiling of protein-associated RNAs are of great importance for functional studies of RNA-binding proteins and gene regulation. Chapter 13 describes the method of isolating ARGONAUTE protein-associated mRNAs and small RNAs using immunoprecipitation in tandem with quantification and profiling. This method can be easily adapted to analyzing RNAs associated with any other RNA-binding proteins.

Chapter 14 describes a powerful method to identify new viruses using RNA-Seq. This approach does not require prior knowledge of viral sequences and is coupled to enrichment of small RNAs and deep high-throughput sequencing that allows the detection of rare viral RNAs (e.g., (11)).

While generating vast amounts of sequence data has become routine and increasingly economical, the bottleneck is in the computational analysis of the data (12). This volume, therefore, includes chapters on bioinformatics methods to digest high-throughput RNA expression data collected by next-generation sequencing (Chaps. 7, 15–17). Readers will notice that these chapters do not necessarily follow the familiar series format mainly developed for wet lab protocols.

Finally, we hope this book provides a comprehensive set of techniques and methods on isolating and analyzing mRNAs, small RNAs, and modified RNAs, which can assist you in your gene expression studies.

References

1. Bonner RF et al. (1997) Laser capture microdissection: molecular analysis of tissue. Science 278(5342):1481–1483.
2. Portillo M et al. (2009) Isolation of RNA from laser-capture-microdissected giant cells at early differentiation stages suitable for differential transcriptome analysis. Mol Plant Pathol 10(4):523–535.
3. Pastore JJ et al. (2011) LATE MERISTEM IDENTITY2 acts together with LEAFY to activate APETALA1. Development 138(15):3189–3198.
4. Lauhon CT (2006) Orchestrating sulfur incorporation into RNA. Nat Chem Biol 2(4):182–183.
5. Chowrira BM, Burke JM (1992) Extensive phosphorothioate substitution yields highly active and nuclease-resistant hairpin ribozymes. Nucleic Acids Res 20(11):2835–2840.
6. Scott EC, Uhlenbeck OC (1999) A re-investigation of the thio effect at the hammerhead cleavage site. Nucleic Acids Res 27(2):479–484.

7. Qiu J, Pintel D (2008) Processing of adeno-associated virus RNA. Front Biosci 2008. 13:3101–3115.
8. Filichkin SA et al. (2010) Genome-wide mapping of alternative splicing in *Arabidopsis thaliana*. Genome Res 20(1):45–58.
9. Velculescu VE et al. (1995) Serial analysis of gene expression. Science 270(5235):484–487.
10. Matsumura H et al. (2010) High-throughput SuperSAGE for digital gene expression analysis of multiple samples using next generation sequencing. PLoS ONE 5(8):e12010.
11. Zhang Y et al. (2011) Association of a novel DNA virus with the grapevine vein-clearing and vine decline syndrome. Phytopathology 101(9):1081–1090.
12. Schuster SC (2008) Next-generation sequencing transforms today's biology. Nat Methods 5(1):16–18.

Contents

Preface... *v*
Contributors... *xi*

1 SuperSAGE: Powerful Serial Analysis of Gene Expression 1
   Hideo Matsumura, Naoya Urasaki, Kentaro Yoshida,
   Detlev H. Krüger, Günter Kahl, and Ryohei Terauchi

2 Northern Blot Analysis for Expression Profiling of mRNAs
   and Small RNAs.. 19
   Ankur R. Bhardwaj, Ritu Pandey, Manu Agarwal,
   and Surekha Katiyar-Agarwal

3 Construction of RNA-Seq Libraries from Large and Microscopic
   Tissues for the Illumina Sequencing Platform 47
   Hagop S. Atamian and Isgouhi Kaloshian

4 Strand-Specific RNA-Seq Applied to Malaria Samples 59
   Nadia Ponts, Duk-Won D. Chung, and Karine G. Le Roch

5 RNA In Situ Hybridization in Arabidopsis............................ 75
   Miin-Feng Wu and Doris Wagner

6 Laser Microdissection of Cells and Isolation of High-Quality
   RNA After Cryosectioning 87
   Marta Barcala, Carmen Fenoll, and Carolina Escobar

7 Detection and Quantification of Alternative
   Splicing Variants Using RNA-seq 97
   Douglas W. Bryant Jr, Henry D. Priest, and Todd C. Mockler

8 Separating and Analyzing Sulfur-Containing RNAs
   with Organomercury Gels.. 111
   Elisa Biondi and Donald H. Burke

9 RNAse Mapping and Quantitation of RNA Isoforms..................... 121
   Lakshminarayan K. Venkatesh, Olufemi Fasina, and David J. Pintel

10 Detection and Quantification of Viral and Satellite RNAs in Plant Hosts 131
   Sun-Jung Kwon, Jang-Kyun Seo, and A.L.N. Rao

11 In Situ Detection of Mature miRNAs in Plants Using
   LNA-Modified DNA Probes.. 143
   Xiaozhen Yao, Hai Huang, and Lin Xu

12 Small RNA Isolation and Library Construction for Expression
   Profiling of Small RNAs from *Neurospora* and *Fusarium*
   Using Illumina High-Throughput Deep Sequencing..................... 155
   Gyungsoon Park and Katherine A. Borkovich

13 Isolation and Profiling of Protein-Associated Small RNAs 165
   Hongwei Zhao, Yifan Lii, Pei Zhu, and Hailing Jin
14 New Virus Discovery by Deep Sequencing of Small RNAs 177
   Kashmir Singh, Ravneet Kaur, and Wenping Qiu
15 Global Assembly of Expressed Sequence Tags 193
   Foo Cheung
16 Computational Analysis of RNA-seq 201
   Scott A. Givan, Christopher A. Bottoms, and William G. Spollen
17 Identification of MicroRNAs and Natural Antisense
   Transcript-Originated Endogenous siRNAs from
   Small-RNA Deep Sequencing Data............................... 221
   Weixiong Zhang, Xuefeng Zhou, Jing Xia, and Xiang Zhou

Index 229

Contributors

Manu Agarwal • *Department of Botany, University of Delhi, Delhi, India*
Hagop S. Atamian • *Department of Nematology, Center for Disease Vector Research, Center for Plant Cell Biology, University of California, Riverside, CA, USA*
Marta Barcala • *Facultad de Ciencias del Medio Ambiente, Universidad de Castilla-La Mancha, Toledo, Spain*
Ankur R. Bhardwaj • *Department of Botany, University of Delhi, Delhi, India*
Elisa Biondi • *University of Missouri-School of Medicine, Department of Molecular Microbiology and Immunology and Department of Biochemistry, Columbia, MO, USA*
Katherine A. Borkovich • *Department of Plant Pathology and Microbiology, Institute for Integrative Genome Biology, University of California, Riverside, CA, USA*
Christopher A. Bottoms • *Informatics Research Core Facility, University of Missouri, Columbia, MO, USA*
Douglas W. Bryant Jr. • *The Donald Danforth Plant Science Center, St. Louis, MO, USA; Intuitive Genomics, Inc., St. Louis, MO, USA*
Donald H. Burke • *University of Missouri School of Medicine, Department of Molecular Microbiology and Immunology and Department of Biochemistry, Columbia, MO, USA*
Foo Cheung • *Center for Human Immunology, Autoimmunity, and Inflammation, National Institute of Health, Bethesda, MD, USA*
Duk-Won D. Chung • *Department of Cell Biology and Neuroscience, Institute for Integrative Genome Biology, Center for Disease Vector Research, University of California, Riverside, CA, USA*
Carolina Escobar • *Facultad de Ciencias del Medio Ambiente, Universidad de Castilla-La Mancha, Toledo, Spain*
Olufemi Fasina • *Department of Molecular Microbiology and Immunology, School of Medicine, University of Missouri-Columbia, Columbia, MO, USA*
Carmen Fenoll • *Facultad de Ciencias del Medio Ambiente, Universidad de Castilla-La Mancha, Toledo, Spain*
Walter Gassmann • *Division of Plant Sciences, Christopher S. Bond Life Sciences Center and Interdisciplinary Plant Group, University of Missouri, Columbia, MO, USA*
Scott A. Givan • *Department of Molecular Microbiology and Immunology, Informatics Research Core Facility, University of Missouri, Columbia, MO, USA*
Hai Huang • *National Laboratory of Plant Molecular Genetics, Shanghai Institute of Plant Physiology and Ecology, Shanghai Institutes for Biological Sciences, Chinese Academy of Sciences, Shanghai, China*
Hailing Jin • *Department of Plant Pathology and Microbiology, Institute for Integrative Genome Biology, University of California, Riverside, CA, USA*
Günter Kahl • *Biocenter, University of Frankfurt am Main, Frankfurt, Germany*

ISGOUHI KALOSHIAN • *Department of Nematology, Center for Disease Vector Research, Center for Plant Cell Biology, University of California, Riverside, CA, USA*

SUREKHA KATIYAR-AGARWAL • *Department of Plant Molecular Biology, University of Delhi South Campus, New Delhi, India*

RAVNEET KAUR • *Center for Grapevine Biotechnology, William H. Darr School of Agriculture, Missouri State University, Mountain Grove, MO, USA; Plant Science Department, McGill University, Ste. Anne de Bellevue, QC, Canada*

DETLEV H. KRÜGER • *Institute of Medical Virology, University Hospital Charité, Berlin, Germany*

SUN-JUNG KWON • *Department of Plant Pathology and Microbiology, University of California, Riverside, CA, USA*

KARINE G. LE ROCH • *Department of Cell Biology and Neuroscience, Institute for Integrative Genome Biology, Center for Disease Vector Research, University of California, Riverside, CA, USA*

YIFAN LII • *Department of Plant Pathology and Microbiology, Center for Plant Cell Biology and Institute for Integrative Genome Biology, University of California, Riverside, CA, USA*

HIDEO MATSUMURA • *Gene Research Center, Shinshu University, Ueda, Nagano, Japan*

TODD C. MOCKLER • *The Donald Danforth Plant Science Center, St. Louis, MO, USA; Division of Biology and Biomedical Sciences, Washington University, St. Louis, MO, USA*

RITU PANDEY • *Department of Plant Molecular Biology, University of Delhi South Campus, New Delhi, India*

GYUNGSOON PARK • *Plasma Bioscience Research Institute, Kwangwoon University, Wolgaedong, Nowongu, Seoul, Republic of Korea*

DAVID J. PINTEL • *Department of Molecular Microbiology and Immunology, School of Medicine, University of Missouri-Columbia, Columbia, MO, USA*

NADIA PONTS • *Department of Cell Biology and Neuroscience, Institute for Integrative Genome Biology, Center for Disease Vector Research, University of California, Riverside, CA, USA*

HENRY D. PRIEST • *The Donald Danforth Plant Science Center, St. Louis, MO, USA; Division of Biology and Biomedical Sciences, Washington University, St. Louis, MO, USA*

WENPING QIU • *Center for Grapevine Biotechnology, William H. Darr School of Agriculture, Missouri State University, Mountain Grove, MO, USA*

A.L.N. RAO • *Department of Plant Pathology and Microbiology, University of California, Riverside, CA, USA*

JANG-KYUN SEO • *Department of Plant Pathology and Microbiology, University of California, Riverside, CA, USA*

KASHMIR SINGH • *Center for Grapevine Biotechnology, William H. Darr School of Agriculture, Missouri State University, Mountain Grove, MO, USA; Department of Biotechnology, Panjab University, Chandigarh, Punjab, India*

WILLIAM G. SPOLLEN • *Informatics Research Core Facility, University of Missouri, Columbia, MO, USA*

RYOHEI TERAUCHI • *Iwate Biotechnology Research Center, Kitakami, Iwate, Japan*

NAOYA URASAKI • *Okinawa Agricultural Research Center, Itoman, Okinawa, Japan*
LAKSHMINARAYAN K. VENKATESH • *Department of Molecular Microbiology and Immunology, School of Medicine, University of Missouri-Columbia, Columbia, MO, USA*
DORIS WAGNER • *Department of Biology, University of Pennsylvania, Philadelphia, PA, USA*
MIIN-FENG WU • *Department of Biology, University of Pennsylvania, Philadelphia, PA, USA*
JING XIA • *Department of Computer Science and Engineering, Washington University, St. Louis, MO, USA*
LIN XU • *National Laboratory of Plant Molecular Genetics, Shanghai Institute of Plant Physiology and Ecology, Shanghai Institutes for Biological Sciences, Chinese Academy of Sciences, Shanghai, China*
XIAOZHEN YAO • *National Laboratory of Plant Molecular Genetics, Shanghai Institute of Plant Physiology and Ecology, Shanghai Institutes for Biological Sciences, Chinese Academy of Sciences, Shanghai, China*
KENTARO YOSHIDA • *Iwate Biotechnology Research Center, Kitakami, Iwate, Japan*
WEIXIONG ZHANG • *Department of Computer Science and Engineering, Fudan University, Shanghai, China; Department of Computer Science and Engineering, Washington University, St. Louis, MO, USA; Department of Genetics, Washington University, St. Louis, MO, USA*
HONGWEI ZHAO • *Department of Plant Pathology and Microbiology, Center for Plant Cell Biology and Institute for Integrative Genome Biology, University of California, Riverside, CA, USA*
XIANG ZHOU • *Department of Computer Science and Engineering, Washington University, St. Louis, MO, USA*
XUEFENG ZHOU • *Department of Computer Science and Engineering, Washington University, St. Louis, MO, USA*
PEI ZHU • *Department of Plant Pathology and Microbiology, Center for Plant Cell Biology and Institute for Integrative Genome Biology, University of California, Riverside, CA, USA*

Chapter 1

SuperSAGE: Powerful Serial Analysis of Gene Expression

Hideo Matsumura, Naoya Urasaki, Kentaro Yoshida, Detlev H. Krüger, Günter Kahl, and Ryohei Terauchi

Abstract

SuperSAGE is a variant of the Serial Analysis of Gene Expression (SAGE) technology, based on counting transcripts by sequencing analysis of short sequence tags. In SuperSAGE, 26 bp tags are extracted from cDNA using the Type III restriction endonuclease EcoP15I. The use of a longer tag size in SuperSAGE allows a secure tag-to-gene annotation in any eukaryotic organism. We have succeeded in combining SuperSAGE and high-throughput sequencing technology (Now- or Next-Generation Sequencing, NGS) in an approach we call High-throughput SuperSAGE (HT-SuperSAGE). This approach allows deep transcriptome analysis and multiplexing, while reducing time, cost, and effort for the analysis. In this chapter, we present the detailed HT-SuperSAGE protocol for both the Illumina Genome Analyzer and also the AppliedBiosystems SOLiD sequencer.

Key words: SuperSAGE, Next-generation sequencing, Transcriptome, Digital gene expression analysis, High-throughput analysis

1. Introduction

Various techniques are now available for the measurement of RNA abundance in cells or tissues. Northern blot analysis and RT-PCR or qPCR (real-time PCR) are the conventional tools for analyzing transcripts originating from individual genes. With the accumulation of whole genome sequence or cDNA (EST) sequence data, large-scale gene expression (or transcriptome) analysis is an effective and common tool in biological studies. Microarrays of various geometry and density were driving high-throughput gene expression profiling for more than a decade (1), and catalyzed transcriptome analysis of many, mostly model organisms. However, it is a closed architecture format, which can detect only the transcription of genes that are spotted on the array. Alternatively, SAGE (Serial

Analysis of Gene Expression) was developed as a technology for large-scale gene expression analysis that is based on DNA sequencing (2). In SAGE, a short "tag" fragment (13–15 bp) is extracted from a defined position of each cDNA, and a large number of tags are pooled and sequenced. By listing the count and annotation of thousands of such tags, one can obtain a comprehensive and quantitative profile of gene expression. SAGE is an open-architecture method, whereby researchers theoretically can address all the expressed transcripts simply by increasing the number of analyzed tags. However, the original SAGE method suffered from the problem of accuracy in tag-to-gene annotation, owing to the short size of the tag. Therefore, improved versions of SAGE were established that obtained longer tags from cDNAs, like LongSAGE (3) employing MmeI to isolate 21 bp tags, and SuperSAGE (4) that generated much longer tags (26 bp). Recent rapid advancements of DNA sequencing technologies, called next- (or now-) generation sequencing (NGS) (5), dramatically improved this SuperSAGE technology. Currently available NGS is based on massively parallel short read sequencing, which perfectly fits to sequence SuperSAGE tags. Therefore, a combination of SuperSAGE and one of the NGS platforms now allows the analysis of millions of tags, multiplexing the analysis of several different samples, and reduces time, cost, and effort (6). We named this new technology High-throughput SuperSAGE (HT-SuperSAGE). Here, we describe the detailed protocol of HT-SuperSAGE for both the Illumina Genome Analyzer and also the AppliedBiosystems SOLiD sequencer, two of the major massively parallel sequencing technology platforms.

2. Materials

2.1. Adapter Preparation

1. Adapter oligonucleotides: adapter oligonucleotide synthesis and end-labeling were done by a supplier. Sequences of IL-adapter-1 and SLD-adapter-1 oligonucleotides are shown in Table 1 (see Note 1), sequences of IL-adapter-2 and SLD-adapter-2 oligonucleotides in Tables 2 and 3, respectively (see Note 2). These oligonucleotides were purified by an Oligonucleotide Purification Cartridge (OPC) (see Note 3).
2. LoTE buffer: 3 mM Tris–HCl, pH 7.5, 0.2 mM EDTA.
3. Polynucleotide kinase buffer (10×): 0.5 M Tris–HCl, pH 8.0, 0.1 M $MgCl_2$, 50 mM DTT.

2.2. cDNA Synthesis

1. First strand buffer (5×, Invitrogen, Carlsbad, CA): 250 mM Tris–HCl, pH 8.0, 375 mM KCl, 15 mM $MgCl_2$.
2. Biotinylated adapter-oligo (dT) primer: Synthesized biotin-labeled oligonucleotides (5′-biotin-CTGATGTAGAGGTAC-

Table 1
Oligonucleotide sequences of IL-adapter-1 and SLD-adapter-1

Adapter oligo name	Oligonucleotide sequences (5′→3′)
IL-adapter-1 sense[a]	ACAGGTTCAGAGTTCTACAGTCCGACGATCWWWW
IL-adapter-1 antisense[a]	NNXXXXGATCGTCGGACTGTAGAACTCTGAACCTGT-amino
SLD-adapter-1 sense[b]	CCACTACGCCTCCGCTTTCCTCTCTATGGGCAGTCGGTGATYYYYYY
SLD-adapter-1 antisense[b]	NNZZZZZZATCACCGACTGCCCATAGAGAGGAAAGCGGAGGCGTAGTGGTT-amino

[a]WWWW encodes 4-bases variable index sequences and should be complementary to XXXX, see Note 1
[b]YYYYYY encodes 6-bases variable index sequences and should be complementary to ZZZZZZ, see Note 2

Table 2
Oligonucleotide sequences of IL-adapter-2 for NlaIII-, DpnII-, or BfaI-digested cDNA

Adapter oligo name	Oligonucleotide sequences (5′→3′)
IL-adapter-2Nla-sense[a]	CAAGCAGAAGACGGCATACGATCTAACGATGTACGCAGCAGCATG
IL-adapter-2Nla-antisense[a]	CTGCTGCGTACATCGTTAGATCGTATGCCGTCTTCTGCTTG-amino
IL-adapter-2Dpn-sense[b]	CAAGCAGAAGACGGCATACGATCTAACGATGTACGCAGCAG
IL-adapter-2Dpn-antisense[b]	GATCCTGCTGCGTACATCGTTAGATCGTATGCCGTCTTCTGCTTG-amino
IL-adapter-2Bfa-sense[c]	CAAGCAGAAGACGGCATACGATCTAACGATGTACGCAGCAGC
IL-adapter-2Bfa-antisense[c]	CTAGCTGCTGCGTACATCGTTAGATCGTATGCCGTCTTCTGCTTG-amino

[a]IL-adapter-2 for ligation to NlaIII-digested cDNA ends
[b]IL-adapter-2 for ligation to DpnII-digested cDNA ends
[c]IL-adapter-2 for ligation to BfaI-digested cDNA ends

CGGATGCCAGCAGTTTTTTTTTTTTTTTTTTTT-3′), HPLC-purified (see Notes 4 and 5), were dissolved in LoTE (1 μg/μL).

3. 0.1 M DTT (dithiothreitol, Invitrogen, Carlsbad, CA).
4. 10 mM dNTP (Invitrogen, Carlsbad, CA): 10 mM each of dATP, dTTP, dCTP, and dGTP.
5. SuperScript II reverse transcriptase (Invitrogen, Carlsbad, CA).

Table 3
Oligonucleotide sequences of SLD-adapter-2 for NlaIII-, DpnII- or TaqI-digested cDNA

Adapter oligo name	Oligonucleotide sequences (5'→3')
SLD-adapter-2Nla-sense[a]	TTCCTCATTCTCTCAAGCAGAAGACGGCATACGAAATGATACGGCGACCACCGACAGGTCTAACGATGTACGCAGCAGCATG
SLD-adapter-2Nla-antisense[a]	CTGCTGCGTACATCGTTAGACCTGTCGGTGGTCGCCGTATCATTTCGTATGCCGTCTTCTGCTTGAGAGAATGAGGAA-amino
SLD-adapter-2Dpn-sense[b]	TTCCTCATTCTCTCAAGCAGAAGACGGCATACGAAATGATACGGCGACCACCGACAGGTCTAACGATGTACGCAGCAG
SLD-adapter-2Dpn-antisense[b]	GATCCTGCTGCGTACATCGTTAGACCTGTCGGTGGTCGCCGTATCATTTCGTATGCCGTCTTCTGCTTGAGAGAATGAGGAA-amino
SLD-adapter-2Taq-sense[c]	TTCCTCATTCTCTCAAGCAGAAGACGGCATACGAAATGATACGGCGACCACCGACAGGTCTAACGATGTACGCAGCAGT
SLD-adapter-2Taq-antisense[c]	TCGACTGCTGCGTACATCGTTAGACCTGTCGGTGGTCGCCGTATCATTTCGTATGCCGTCTTCTGCTTGAGAGAATGAGGAA-amino

[a]SLD-adapter-2 for ligation to NlaIII-digested cDNA ends
[b]SLD-adapter-2 for ligation to DpnII-digested cDNA ends
[c]SLD-adapter-2 for ligation to TaqI-digested cDNA ends

6. Second strand buffer (Invitrogen, Carlsbad, CA): 100 mM Tris–HCl (pH 6.9), 450 mM KCl, 23 mM $MgCl_2$, 0.75 mM β-NAD^+, 50 mM $(NH4)_2SO_4$.
7. *Escherichia coli* DNA polymerase (10 U/μL, Invitrogen, Carlsbad, CA).
8. *E. coli* DNA ligase (1.2 U/μL, Invitrogen, Carlsbad, CA).
9. *E. coli* RNase H (2 U/μL, Invitrogen, Carlsbad, CA).
10. Binding buffer (PB buffer) in Qiaquick PCR purification kit (Qiagen, Germany).
11. Qiaquick spin column in Qiaquick PCR purification kit (Qiagen, Germany).
12. Washing buffer (PE buffer, 5×) (Qiagen, Germany): prepare 1× solution by adding ethanol before use.

2.3. Tag Extraction

1. NlaIII (10 U/μL, New England Biolab, Ipswich, MA): Store at –70°C.
2. DpnII (10 U/μL, New England Biolab, Ipswich, MA): Store at –20°C.

3. BfaI (5 U/μL, New England Biolab, Ipswich, MA): Store at −20°C.
4. TaqI (20 U/μL, New England Biolab, Ipswich, MA): Store at −20°C.
5. NEBuffer 4 (10×, New England Biolabs, Ipswich, MA): 20 mM Tris–acetate, pH 7.9, 50 mM potassium acetate, 10 mM magnesium acetate, 1 mM DTT.
6. NEBuffer 3 (10×, New England Biolabs, Ipswich, MA): 50 mM Tris–HCl, pH 7.9, 100 mM NaCl, 10 mM $MgCl_2$, 1 mM DTT.
7. BSA (10 mg/mL, New England Biolabs, Ipswich, MA).
8. Streptavidin-coated magnetic beads (Dynabeads M-270 Streptavidin, see Note 6) (10 mg/mL, Invitrogen, CA): Store at 4°C.
9. Siliconized microtube (1.5 mL).
10. Binding and washing buffer (B&W buffer) (2×): 10 mM Tris–HCl, pH 7.5, 1 mM EDTA, 2 M NaCl.
11. T4 DNA ligase (2,000 U/μL): Store at −20°C.
12. T4 DNA ligase buffer (5×): 250 mM Tris–HCl, pH 7.5, 50 mM $MgCl_2$, 5 mM ATP, 50 mM DTT, 125 μg/mL BSA.
13. EcoP15I (10,000 U/μL, New England Biolabs, Ipswich, MA): Store at −20°C.
14. 10× ATP solution (1 mM, New England Biolabs, Ipswich, MA).
15. 100× BSA solution (100 μg/mL, New England Biolabs, Ipswich, MA).
16. Phenol:chloroform:isoamyl alcohol (25:24:1). Store at 4°C.
17. Ammonium acetate: 10 M solution.
18. Glycogen solution (20 mg/mL).

2.4. Indexed-Adapter Ligation and PCR

1. Phusion HF 5× Buffer (FINZYMES, Finland).
2. dNTP solution: 10 mM each of dATP, dTTP, dCTP, and dGTP.
3. $MgCl_2$ solution (50 mM, FINZYMES, Finland).
4. IL-adapter-1 primer: 5′-AATGATACGGCGACCACCGACA GGTTCAGAGTTCTACAGTCCGA-3′.

   IL-adapter-2 primer: 5′-CAAGCAGAAGACGGCATACGA-3′.

   SLD-adapter-1 primer: 5′- CCACTACGCCTCCGCTTTCCT CTC -3′.

   SLD-adapter-2 primer: 5′-CTGCCCCGGGTTCCTCATTCT CTCAAGCAGAAGA-3′.

These oligonucleotides are synthesized and purified by the OPC and are dissolved in LoTE to a final concentration of 100 pmol/μL.

5. Phusion Hot Start DNA polymerase (2 U/μL, FINZYMES, Finland): Store at –20°C.
6. Acrylamide/Bisacrylamide solution (40%, 19:1): Store at 4°C.
7. *N*, *N*, *N*, *N'*-Tetramethyl-ethylenediamine (TEMED). Store at 4°C.
8. Ammonium persulfate: prepare 10% solution in sterilized water and store at 4°C.
9. 6× Loading dye: 30%(v/v) glycerol, 0.25% (w/v) bromophenol blue and 0.25% (w/v) xylene cyanol.
10. SYBR green solution (Molecular Probe, Eugene, OR): Original SYBR green stock solution was diluted 10,000 times with 1× TAE buffer. Store at 4°C.
11. 20 bp DNA marker ladder (200 ng/μL).

2.5. Purification of PCR Product

1. PB buffer in MinElute Reaction Cleanup kit (Qiagen, Germany).
2. MinElute spin column in MinElute Reaction Cleanup kit (Qiagen, Germany).
3. Spin-X column (Corning, Corning, NY).

2.6. Multiplexing DNA Samples for Sequencing Analysis

1. Agilent 2100 Bioanalyzer (Agilent Technologies, Wilmington, DE).
2. Agilent DNA 1000 kit includes chips and Gel-Dye mix (Agilent Technologies, Wilmington, DE).

3. Method

In HT-SuperSAGE using NGS technologies, 26 bp tags were extracted from cDNA the same way as described in the original SuperSAGE procedure. However, instead of ditags in the original SuperSAGE method, single tags, sandwiched between two adapters, were prepared. These single tags with adapters were once PCR amplified and directly subjected to sequencing without plasmid vector cloning. Experimental procedures for SuperSAGE were almost identical for either one of the two massively parallel sequencing technologies Illumina GA or ABI SOLiD, although specific adapters are used for each sequencing technology. For analyzing multiple SuperSAGE libraries in a single run of sequencing, adapter fragments harboring different index sequences are ligated to the tags derived from different biological samples. After sequencing of pooled indexed fragments (adapter-tags), the obtained sequence reads are bioinformatically separated according to their index sequences. Employing this procedure, the transcriptome of multiple samples could be analyzed simultaneously. It has already been demonstrated that tag sequences from independent samples could be accurately discriminated by using a 4-bp index sequence (6). In the present protocol, we also employed a 6-bp index in the adapter for SOLiD sequencing, which is tolerant to possible failure in sample discrimination by single-base errors in the index sequences (Fig. 1; Table 4). Additionally, SuperSAGE tags can be extracted from different positions in cDNA sequences by changing the four-base cutter anchoring enzyme for cDNA digestion (Fig. 1). We also describe a procedure for the use of several different anchoring enzymes in the present protocol.

3.1. Adapter Preparation

1. Dissolve synthesized adapter oligonucleotides in LoTE buffer (100 pmol/μL). Appropriate adapter-2 (IL-adapter-2 or SLD-adapter-2) for each anchoring enzyme should be prepared (see Note 2). The variety of different indexed IL-adapter-1

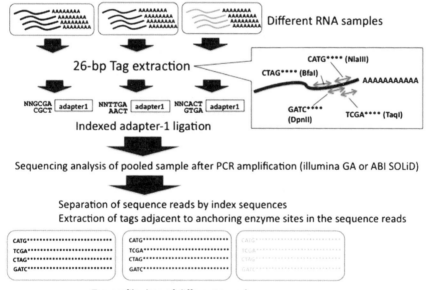

Fig. 1. Scheme of high-throughput SuperSAGE analysis. RNAs from different tissues are extracted, and 26-bp tags recovered from their cDNAs. By changing the anchoring enzyme (NlaIII, DpnII, BfaI, or TaqI in this figure), tags are extracted from different positions in the cDNA. Adapters with different index sequences are designed and ligated to tags from different samples. PCR products of adapter-1 ligated tags are pooled and directly sequenced. Sequence reads are separated by index sequences, and 26 bp tags are extracted bioinformatically from anchoring enzyme sites in the sequence reads.

Table 4
Summary of a SuperSAGE analysis of rice leaf tissues using SOLiD sequencing

		Tag count			
Sample name[a]	Index sequences[b]	NlaIII (CATG)	DpnII (GATC)	TaqI (ACGT)	Total count (in each sample)
Sasanishiki leaf blade (28°C)	AAACCA	123,651	144,132	86,888	354,671
Hitomebore leaf blade (28°C)	AATCCT	436,562	137,362	19,011	592,935
Sasanishiki leaf sheath (28°C)	AACCCC	226,799	191,899	187,816	606,514
Hitomebore leaf sheath (28°C)	AAGCCG	114,834	305,524	108,428	528,786
Sasanishiki leaf blade (15°C)	ATACGA	195,449	144,004	326,349	665,802
Hitomebore leaf blade (15°C)	ATTCGT	153,109	175,312	102,621	431,042
Sasanishiki leaf sheath (15°C)	ATCCGC	149,948	161,584	130,974	442,506
Hitomebore leaf sheath (15°C)	ATGCGG	317,224	110,847	189,316	617,387
Total count (tag from each anchoring enzyme)		1,717,576	1,370,664	1,151,403	4,239,643

[a]One-month-old rice (c.v. Sasanishiki or Hitomebore) plants were kept at different temperatures (28 or 15°C, respectively) for 1 day, and RNAs extracted from leaf blades or leaf sheath tissue of these plants and applied to SuperSAGE
[b]Six-base index was employed in this analysis

or SLD-adapter-1 oligonucleotides depends on the number of independent samples to be multiplexed in a single run of sequencing (see Note 7).

2. Combine 10 μL of each complementary oligonucleotide solution (sense and antisense in each adapter oligonucleotide) in a tube. Add 3 μL 10× polynucleotide kinase buffer and 7 μL LoTE.

3. Each sample mixture with a total volume of 30 μL is denatured by incubating at 95°C for 2 min and cooled down to 20°C for annealing complementary oligonucleotides. The annealed double-stranded DNA is designated "adapter" (e.g., IL-adapter-1, SLD-adapter-2Nla, etc.).

3.2. cDNA Synthesis

1. The synthesis of double-stranded cDNA follows the protocol described in the SuperScriptII double-strand cDNA synthesis kit (Invitrogen). Total RNA (2–10 μg) is dissolved in 11 μL DEPC-treated water and incubated at 70°C for 10 min after adding 1 μL biotinylated adapter-oligo dT primer (100 pmol). Denatured RNA solution is immediately placed on ice, and 4 μL 5× First Strand buffer, 2 μL 0.1 M DTT, 1 μL 10 mM dNTP and 1 μL SuperScriptII reverse transcriptase are added for first-strand cDNA synthesis. The reaction mixture is incubated at 45°C for 1 h.

2. For second-strand cDNA synthesis, 30 μL 5× Second Strand buffer, 91 μL sterile water, 3 μL 10 mM dNTP, 4 μL *E. coli* DNA polymerase, 1 μL *E. coli* RNase H, and 1 μL *E. coli* DNA ligase are added to 20 μL first-strand cDNA solution, and mixed. Incubate at 16°C for 2 h.

3. For purification of the synthesized double-stranded cDNA, 750 μL PB buffer is added to the cDNA solution, and the mixed solution applied to a Qiaquick spin column of the same kit and centrifuged at 10,000×*g* for 1 min. After discarding the flow-through, 750 μL washing buffer (1× PE buffer; prepared as described in Subheading 2.2) are added to the column. Centrifuge at 10,000×*g* for 1 min and discard flow-through. For completely drying, the column is centrifuged at maximum speed for 1 min. After the column is transferred to a new 1.5-mL microtube, 30 μL LoTE is added for elution. The eluate (purified cDNA) is collected by centrifugation at 10,000×*g* for 1 min.

3.3. Tag Extraction

1. Purified double-stranded cDNA is digested with the anchoring enzyme (four-base cutter restriction enzyme like NlaIII, DpnII, BfaI, or TaqI, see Note 8).

   For digestion of cDNA with NlaIII, 20 μL NEBuffer 4, 2 μL BSA, 143 μL LoTE, 5 μL NlaIII (10 U/μL) are added to the cDNA solution, mixed, and incubated at 37°C for 1.5 h.

   For digestion of cDNA with DpnII, 20 μL NEBuffer 3, 145 μL LoTE, 5 μL DpnII (10 U/μL) are added to the cDNA solution, mixed, and incubated at 37°C for 1.5 h.

   For digestion of cDNA with BfaI, 20 μL NEBuffer 4, 145 μL LoTE, 5 μL BfaI (5 U/μL) are added to the cDNA solution, mixed, and incubated at 37°C for 1.5 h.

   For digestion of cDNA with TaqI, 20 μL NEBuffer 4, 2 μL BSA, 143 μL LoTE, 5 μL TaqI (20 U/μL) are added to the cDNA solution, mixed, and incubated at 65°C for 1.5 h.

2. Prepare 100 μL of a suspension of streptavidin-coated magnetic beads (Dynabeads M-270) in a siliconized 1.5-mL microtube (see Note 9). Place the tubes containing magnetic beads on a magnetic stand and remove the supernatant with a pipette. For washing the magnetic beads, 200 μL of 1× B&W solution is added and beads are suspended well by pipetting. Place the tube on a magnetic stand, and remove and discard the supernatant (see Note 10).

3. To the washed magnetic beads, 200 μL of 2× B&W solution and 200 μL of digested cDNA solution are added and suspended well by pipetting. Leave the tube for 15–20 min at room temperature with occasional mixing, so that the biotinylated cDNAs bind to streptavidin on the magnetic beads.

After digested cDNAs are associated with the beads, the tube is placed on the magnetic stand, and the supernatant is discarded. Magnetic beads are washed three times with 200 μL 1× B&W and once with 200 μL LoTE.

4. For GA sequencing, the appropriate IL-adapter-2 for the employed anchoring enzyme (NlaIII, DpnII, or BfaI) is ligated to the digested cDNAs on the beads (see Note 2). For SOLiD sequencing, the appropriate SLD-adapter-2 for the employed anchoring enzyme (NlaIII, DpnII, or TaqI) is ligated to the digested cDNAs on the beads (see Note 2). To the washed beads, 21 μL LoTE, 6 μL 5× T4 DNA ligase buffer, and 1 μL of either IL-adapter-2 or SLD-adapter-2 solution are added. After mixing buffer and adapter solution, the bead suspension is incubated at 50°C for 2 min. Then the tube is cooled down at room temperature for 15 min and 2 μL T4 DNA ligase (10 U) is added. It is then incubated at 16°C for 2 h.

5. The beads are washed four times with 1× B&W, and three times with LoTE after the ligation reaction. The beads are suspended in 75 μL LoTE.

6. For EcoP15I digestion of the fragments on the magnetic beads, 10 μL 10× NEbuffer 3, 10 μL 10× ATP solution, 1 μL 100× BSA, and 4 μL EcoP15I (10,000 U/μL) are added to the suspended magnetic beads. Incubate the tube at 37°C for 2 h with occasional mixing (see Note 11).

7. After EcoP15I digestion, the bead suspension is placed on the magnetic stand, and the supernatant is collected into a new tube. The beads are re-suspended in 100 μL 1× B&W. After separation on the magnetic stand, the supernatants are retrieved and combined to the previously collected solution (see Note 12) in each tube.

8. To the collected solution, containing adapter-tag fragments, half a volume of phenol:chloroform:isoamyl alcohol (195 μL) is added, shortly vortexed, and spun at $10,000 \times g$ for a few minutes. The upper aqueous layer is transferred to a new tube (see Note 13). For ethanol precipitation, 100 μL 10 M ammonium acetate, 3 μL glycogen, and 900 μL cold ethanol are added to the collected solution (approximately 200 μL). Keep the tube at −80°C for 1 h, and centrifuge at maximum speed for 40 min at 4°C. The resulting pellet is washed twice with 70% ethanol, and dried. Precipitated adapter-2 ligated 26-bp-tag fragments are dissolved in 10 μL LoTE.

3.4. Indexed-Adapter Ligation and PCR

1. Prepare adapter-1 (IL-adapter-1 for Illumina GA sequencing and SLD-adapter-1 for ABI SOLiD sequencing) with defined index sequences assigned to individual samples (see Note 1). For the ligation reaction, 3 μL 5× T4 DNA ligase buffer and 0.5 μL adapter solution are added to the solution of the

adapter-2 ligated tags, which were released from the beads by EcoP15I digestion and purified. Incubate the tube at 50°C for 2 min, and subsequently keep it at room temperature for 15 min. After the tubes have cooled down, 1.5 μL T4 DNA ligase (7.5 U) are added, and incubated at 16°C for 2 h (see Note 14).

2. For PCR amplification of adapter-ligated tags, a PCR mixture containing 3 μL 5× Phusion HF buffer, 0.3 μL 2.5 mM dNTP, 0.1 μL 50 mM $MgCl_2$, 0.15 μL of each primer (IL-adapter-1 primer and IL-adapter-2 primer for GA sequencing, or SLD-adapter-1 primer and SLD-adapter-2 primer for SOLiD sequencing), 10.1 μL distilled water, 1 μL ligation solution, and 0.2 μL Phusion Hot Start DNA polymerase is prepared in a tube (see Note 15).

3. PCR cycling: 98°C for 2 min, then ten cycles each at 98°C for 30 s, and 60°C for 30 s (see Note 16).

4. The size of the amplified PCR product is confirmed by polyacrylamide gel electrophoresis (PAGE, see Note 17). Prepare an 8% PAGE gel by mixing 3.5 mL 40% acrylamide/bisacrylamide solution, 13.5 mL distilled water, 350 μL 50× TAE buffer, 175 μL 10% ammonium persulfate, and 15 μL TEMED. Pour the solution onto the gel plate (12 × 12 cm, 1 mm thickness), and insert a comb (no stacking gel).

5. Running buffer (1× TAE) is prepared and added to the upper and lower electrophoresis chambers. Then, 3 μL 6× loading dye is added to 15 μL of the PCR solution, and loaded into the well. Two microliters of a 20-bp ladder is also loaded as molecular size marker. Run the gel at 75 V for 10 min, and then at 150 V for around 30 min (until the BPB dye front has migrated two-thirds down the gel).

6. After electrophoresis, the gel is removed from the plate. Pour 1 mL SYBR green solution (diluted in 1× TAE buffer) onto the plastic wrap, and place the gel on it. Further, disperse 1 mL SYBR green solution onto the gel. After a 2-min staining period, the gel is placed on a UV transilluminator. The size of the expected amplified fragment (tags sandwiched with two adapters) is 125 bp in the sample for GA sequencing (Fig. 2a, see Note 18), and 160 bp for SOLiD sequencing (Fig. 2b, see Note 19).

7. After confirmation of PCR amplification of adapter-ligated tag fragments, repeat PCRs under the same conditions for each sample and adaptor combination in eight tubes.

3.5. Purification of PCR Product

1. After PCR, solutions from all the tubes are collected in a 1.5-mL tube, and 400 μL of ERC buffer supplied with the MinElute Reaction Cleanup kit is added. Prepare a MinElute spin column from the same kit, and transfer a mixture of PCR solution and ERC buffer to the column. Centrifuge at $10,000 \times g$ for 1 min,

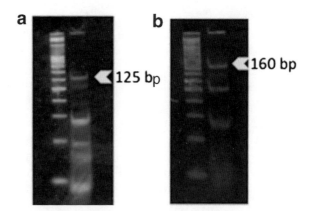

Fig. 2. Size of PCR products of adapter-ligated tag fragments. PCR amplified products of adapter-1 and adapter-2 ligated tags are run on 8% polyacrylamide gels and stained with SYBR green. The PCR product for Illumina GA sequencing (**a**) and ABI SOLiD sequencing (**b**) is loaded on the *right lane* in each panel. A 20-bp molecular size marker is run in the *left lane* of each panel. Expected bands of amplified adapter-1-tag-adapter-2 fragments are indicated by *arrow heads* (125 bp in panel **a** and 160 bp in panel **b**).

and discard flow through. Add another 750 µL of washing buffer (PE buffer, ethanol added) to the column. Centrifuge at 10,000 ×g for 1 min and discard flow through. For completely drying the columns, centrifuge at maximum speed for 1 min. After columns are transferred to new 1.5-mL microtubes, 15 µL LoTE is added to the column for elution. Leave the column for 1 min after adding LoTE, centrifuge at 10,000 ×g for 1 min, and collect eluate.

2. Prepare 8% polyacrylamide gel as described in Subheading 3.4. Add 3 µL 6× loading buffer to column-purified PCR product and load it in the well (see Note 20). After running the gel as described in Subheading 3.4, it is stained with SYBR green, and bands are visualized under UV light.

3. Only the 125 bp band for GA sequencing, or the 160 bp band for SOLiD sequencing, is cut out from the gel with a knife and transferred to 0.5-mL microtubes (Fig. 2, see Note 21). Puncture holes at the top and the bottom of the tube with a needle, and place it into a 1.5-mL tube. Centrifuge the tube at maximum speed for 2–3 min. Polyacrylamide gel pieces are collected at the bottom of the tube. Add 300 µL LoTE to the gel fragments and suspend.

4. After incubation at 37°C for 2 h, the gel suspension is transferred to a Spin-X column and centrifuged at maximum speed for 2 min. The eluate is once extracted by phenol/chloroform, and precipitated by adding 100 µL 10 M ammonium acetate, 3 µL glycogen, and 950 µL cold ethanol. Keep it at −80°C for 1 h, and centrifuge at 15,000 ×g for 40 min at 4°C. Wash once with 70% ethanol and dry. The resulting pellet is then dissolved in 10–15 µL LoTE.

3.6. Multiplexing DNA Samples for Sequencing Analysis

1. For quantification of the purified PCR product, it is analyzed with an Agilent Bioanalyzer system (see Note 22). A DNA chip from Agilent DNA 1000 kit is prepared and filled with Gel-Dye Mix supplied with the kit. Load 1 µL purified PCR product in the well of the chip and run the chip in the Agilent 2100 Bioanalyzer.

2. The DNA concentration of the 125 bp or 160 bp fragment, respectively, is measured with 2100 Expert software (Agilent Technologies). Based on this quantification, an equal amount of DNA (PCR product) from each sample is combined in a tube, and the mixture (see Note 23) is sequenced on the Illumina Genome Analyzer or the ABI SOLiD platform.

3.7. Sequence Data Analysis

1. For the extraction of 26 bp tags from the sequence reads and the estimation of tag frequency, in-house programs written in Perl script are used (SuperSAGE_tag_extract_pipe) (6) (see Note 24), and its process is shown in Fig. 3.

2. First, all sequence reads are separated into independent files by the first 4- or 6-base index sequence in each read. From separated

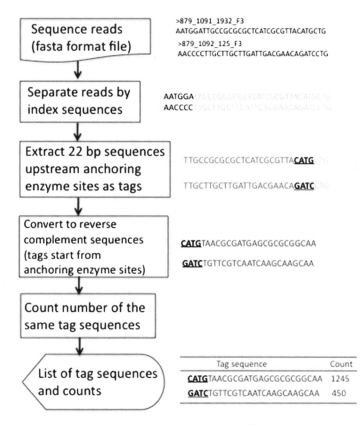

Fig. 3. Flowchart of sequence data analysis in HT-SuperSAGE. Process for sample discrimination by index and tag extraction from sequence reads is illustrated as a flowchart. Our program analyzes files of sequence reads in fasta format as an input, and generates a list of tag sequences and counts as shown in Table 5.

Table 5
An example data of SuperSAGE tag sequences, count, and annotation[a]

Tag sequence[b]	Tag count				Accession number[c]	Gene[d]
	Sasanishiki leaf blade (28°C)	Sasanishiki leaf blade (15°C)	Hitomebore leaf blade (28°C)	Hitomebore leaf blade (15°C)		
CATGTTCGGC TGCACCGAT GCCACCC	4,855	1,701	6,424	3,882	AY445627	Ribulose-1,5-bisphosphate carboxylase/oxygenase small subunit
CATGGACGAG GACGATGCT CCCCCGG	343	343	997	1,118	X67711	Heat shock protein 70
CATGGATGGA CATCGCCGG GGAGTGT	258	24	684	0	DQ872158	(E)-beta-caryophyllene/beta-elemene synthase
CATGGTGAAG GTCATCGCC TGGTACG	98	2	260	16	NM_001 059519	Glyceraldehyde-3-phosphate dehydrogenase A
CATGGGCGTG TGCAGCTGA ATAAGAG	3	1,171	18	1,036	NM_001 187007	Hypothetical protein
CATGCAAACTT AAGGGTAGT GTAGCA	3	376	2	99	NM_001 055931	Monosaccharide transporter 4

[a]Individual tags are selected from SuperSAGE results shown in Table 4
[b]Tags derived from NlaIII sites are shown
[c]Accession numbers correspond to rice (*Oryza sativa* L.) cDNA sequences in GenBank or EMBL, which perfectly matched 26 bp tag sequences
[d]These are genes represented by cDNAs corresponding to the tags

sequence reads in each file, extract 22-base sequences upstream of the respective anchoring enzyme recognition site (NlaIII, DpnII, BfaI, or TaqI).

3. Subsequently, count redundant tag sequences in each file and construct a list of independent tag sequences and their counts. As an example, a summary of analyzed data following the present protocol using SOLiD sequencing is shown in Table 4, and sequences, counts and annotations of selected tags are shown in Table 5.

4. Notes

1. L-adapter-1 or SLD-adapter-1 each contains an address site for the sequencing primer, followed by an index sequence for the discrimination of different samples. In these adapter-1 antisense oligonucleotides, two bases at the 5′-end are synthesized from a mixture of deoxynucleotides ("NN") due to ligation to various two-base 5′-protrusion ends in EcoP15I-digested fragments (adapter-2-ligated tag fragments).

2. IL-adapter-2 or SLD-adapter-2 each contains a recognition site for EcoP15I, and their ends should therefore be compatible with the end of fragments digested with the anchoring enzyme. In Tables 2 and 3, sequences of compatible adapters with NlaIII, DpnII, BfaI, or TaqI sites, respectively, are shown.

3. Incorrect ligation of adapters to tags was prevented by the amino-modification of the 3′-ends of antisense oligonucleotides in all the adapters.

4. Anchoring enzyme sites should not be contained in this biotinylated adapter-oligo (dT) primer.

5. An EcoP15I-recognition site (5′-CAGCAG-3′) is contained in the biotinylated adapter-oligo dT primer.

6. Dynabeads M-270 Streptavidin are paramagnetic beads with hydrophilic surfaces, which bind nonspecific DNA fragments less strongly.

7. Different adapters are prepared in accordance with the number of samples to be analyzed. More than 500 million tags in GA (GAIIx) and more than 700 million tags in SOLiD (SOLiD4) are obtained in a single run of sequencing. The number of tags per sample is inversely related to the number of multiplexed samples in a single sequencing run (e.g., five million tags per sample are expected on average when 100 samples with differently indexed adapters are multiplexed in a single sequencing run of GA).

8. Here, the protocol for NlaIII, DpnII, BfaI, or TaqI is described. We have experience using these four enzymes for SuperSAGE analysis (6) (Table 4). Other four-base cutters are also applicable as anchoring enzymes. BfaI might not be a suiTable anchoring enzyme, since the frequency of its recognition site in cDNA sequences is lower than that of NlaIII, DpnII, or TaqI (6).

9. Employing paramagnetic beads with hydrophilic surfaces and siliconized microtubes, most of the unligated adapters can be eliminated by washing with buffers.

10. Washing of magnetic beads in other steps also follows this procedure.
11. In the adapter-2 ligated cDNA fragments captured on magnetic beads, two EcoP15I sites are present. The enzyme sometimes recognizes the site adjacent to the poly-A tract, and cuts there, which leads to the release of fragments longer than adapter-tags.
12. By repeated washing of the magnetic beads, residual adapter-tag fragments can be collected.
13. Phenol/chloroform extraction in other steps also follows this procedure.
14. Tubes can be incubated for longer periods (e.g., overnight).
15. When preparing the PCR mixture, care should be taken to avoid contamination with previously amplified PCR products. Use separate pipettes and solutions, including water, from those used in the experiments after PCR. Also, use separate labware and gloves.
16. We have demonstrated that ten PCR cycles did not cause an amplification bias, which may affect transcript profiles (6). Even when the number of PCR cycles was increased to 15, no critical amplification bias or distortion in tag profiles could be observed (data not shown).
17. PAGE is better than agarose gel electrophoresis for a good separation of DNA fragments between 100 and 200 bp.
18. The size of PCR products from IL-adapter-1 dimer and IL-adapter-2 dimer is 108 and 180 bp, respectively, whereas the PCR product from IL-adapter-1 and IL-adapter-2 ligates measures around 98 bp.
19. The sizes of PCR products from SLD-adapter-1 dimer and SLD-adapter-2 dimer are 92 and 85 bp, respectively; the size of the PCR product from SLD-adapter-1 and SLD-adapter-2 ligates is about 137 bp.
20. Loading too much DNA to the wells reduces the resolution of PAGE. Normally, purified PCR products from >8 reactions are loaded into two wells separately.
21. Any contamination with DNA fragments of inappropriate size (larger or smaller than 125 or 160 bp, respectively) interferes with the recovery of SuperSAGE tags from sequence data.
22. It is quite important to accurately measure the DNA amount in all samples to obtain the expected number of sequencing reads in each sample after multiplexing. For quantification of DNA, we recommend either the bioanalyzer or fluorescent

(SYBR green) detection method rather than spectrophotometric measurement.

23. If a PCR product is used for SuperSAGE analysis, we consider a total amount of 80 ng as sufficient.

24. For SOLiD sequencing data, csfasta format data are once converted to fasta format data.

Acknowledgments

H.M. is supported by the Program for Promotion of Basic and Applied Researches for Innovations in Bio-oriented Industry (BRAIN). This work is also supported by JSPS grant no. 22380009.

References

1. Schena M, Shalon D, Davis RW, Brown PQ (1995) Quantitative monitoring of gene expression patterns with a complementary DNA microarray. Science 270:467–470
2. Velculescu VE, Zhang L, Vogelstein B, Kinzler KW (1995) Serial analysis of gene expression. Science 270:484–487
3. Saha S, Sparks AB, Rago C, Akmaev V, Wang CJ, Vogelstein B, Kinzler KW, Velculescu VE (2002) Using the transcriptome to annotate the genome. Nat Biotechnol 20:508–512
4. Matsumura H, Reich S, Ito A, Saitoh H, Kamoun S, Winter P, Kahl G, Reuter M, Krüger DH, Terauchi R (2003) Gene expression analysis of host-pathogen interactions by SuperSAGE. Proc Natl Acad Sci USA 100: 15718–15723
5. Metzker ML (2010) Sequencing technologies—the next generation. Nat Rev Genet 11: 31–46
6. Matsumura H, Yoshida K, Luo S, Kimura E, Fujibe T, Albertyn Z, Barrero RA, Krüger DH, Kahl G, Schroth GP, Terauchi R (2010) High-throughput SuperSAGE for digital gene expression analysis of multiple samples using next generation sequencing. PLoS One 5:e1201

Chapter 2

Northern Blot Analysis for Expression Profiling of mRNAs and Small RNAs

Ankur R. Bhardwaj, Ritu Pandey, Manu Agarwal, and Surekha Katiyar-Agarwal

Abstract

Northern analysis is a conventional but gold standard method for detection and quantification of gene expression changes. It not only detects the presence of a transcript but also indicates size and relative comparison of transcript abundance on a single membrane. In recent years, it has been aptly adapted to validate and study the size and expression of small noncoding RNAs. Here, we describe protocols employed in our laboratory for conventional northern analysis with total RNA/mRNA to study gene expression and validation of small noncoding RNAs using low molecular weight fraction of RNAs.

Key words: RNA, Small RNAs, Northern blotting, Gene expression, Quantification

1. Introduction

Besides other approaches, one of the approaches that shed light on the gene functionality is by interrogation of its expression profile at various developmental stages and in different biotic and abiotic stresses. The expression and the steady-state levels of many gene transcripts are in turn regulated by another RNA molecule that either binds to the gene promoter or to the mature mRNA to control the expression and stability of the transcript. The ever increasing discoveries of new genes and the "candidate" small regulatory RNAs at overwhelming rate using high-throughput deep-parallel sequencing has necessitated the standardization of methods that can reliably detect and accurately quantitate their expression levels (1–11). The northern blot technique is one of the most reliable and widely used standard method for validating and quantitating mRNAs and small RNAs (1–7, 12–14). Moreover, the simplicity of

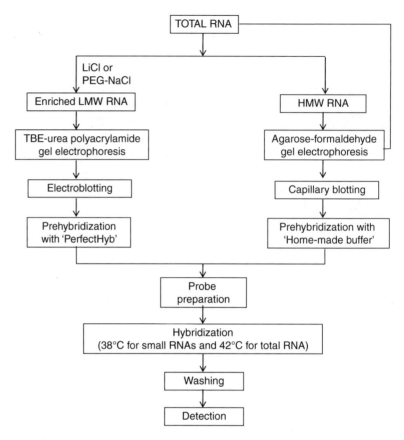

Fig. 1. Schematic representation of northern blot analysis of total RNA as well as small RNAs.

conducting the northern analysis experimentally as well as its economic feasibility has globally propelled this technique as the method of choice to study gene expression changes. Briefly, a standard northern protocol involves size separation via electrophoresis in a polyacrylamide or a denaturing agarose gel, followed by its transfer to a membrane either using electroblotting or via capillary transfer. The RNA is then irreversibly immobilized (also known as crosslinking) to the membrane either by baking the membrane or irradiating it with ultra violet (UV) radiations or by chemical treatments. Subsequently, the membrane is then hybridized with a DNA or a RNA probe (labeled isotopically or nonisotopically). An outline of northern analysis of total RNA and small RNA is depicted in Fig. 1.

Albeit the basic procedure of northern blotting has more or less remained similar over years, subtle but significant improvements in RNA immobilization to the membrane and probe labeling seems to have markedly increased the sensitivity of this technique. Traditionally, nucleic acids are immobilized on membranous solid support by incubation of membrane at 80°C for 1 h and/or ultra violet (UV) crosslinking. In case of small RNAs, 1-ethyl-3-(3-dimethylaminopropyl)-carbodiimide (EDC) mediated crosslinking has been reported to detect those low abundance small RNAs that were barely detectable when the RNAs were UV-crosslinked (15). The UV radiations crosslink the free amino groups of nylon membrane with the RNA nucleotides, predominantly with uracil, thereby posing steric hindrance in accessibility of the immobilized RNA. This problem is further compounded due to the size of small RNAs. Moreover, UV crosslinking of RNA to the membrane does not happen across the entire length and varies with the sequence. However, EDC crosslinking occurs between the free amino groups of the membrane and the 5′ phosphates of the bases. This results in increased accessibility of RNA (due to reduced steric hindrance) and fairly uniform crosslinking (15, 16).

The conventional probe labeling method is based on incorporation of radioisotopes by random priming or nick translation for detection of specific transcripts on a total RNA/mRNA blot and 5′-end labeling of DNA oligonucleotides (by a kinase reaction and γATP) for detection of small RNAs. Isotopic labeling is more widely used than nonisotopic labeling as it produces a better signal-to-noise ratio, thereby facilitating the detection of even moderately less abundant transcripts and small RNAs. However, radioactive method of labeling also has pitfalls. First, handling and disposal of radioactive material requires dedicated instrumentations and laboratories. Second, usage of radioisotopes requires strict regulations and specialized training, and, third radiolabeled probes undergo radiodecay and cannot be stored for longer durations. Moreover, the best signals are obtained if the radiolabeled probes are used for hybridization in the early phases of first half-life of the isotope. On the other hand, though digoxigenin (DIG)-based nonisotopically labeled probes are fairly stable and can be stored for extended durations, their sensitivity for detecting especially small RNAs is far less than the radiolabeled oligonucleotides. Therefore, most of the small RNA northern detections are mostly carried out by using radiolabeled oligonucleotide probes (1, 3, 14, 17–19). A recent improvement that has markedly increased the sensitivity of small RNA detection is the use of locked nucleic acid (LNA) oligonucleotide probes instead of DNA oligonucleotides. LNA bases have 2′-O, 4′-C methylene bridge in the ribose moiety of nucleotide and this modification stabilizes the conformation of the sugar group, thereby increasing the thermal stability of LNA-modified oligonucleotides. This results in high affinity of RNA-probe heteroduplex thereby improving the sensitivity of detecting small RNAs (20, 21).

Next-generation sequencing technologies have enabled high-throughput sequence identification of genes and small noncoding RNAs in various organisms at an unprecedented pace. It is therefore imperative to design validation strategies that match the pace of

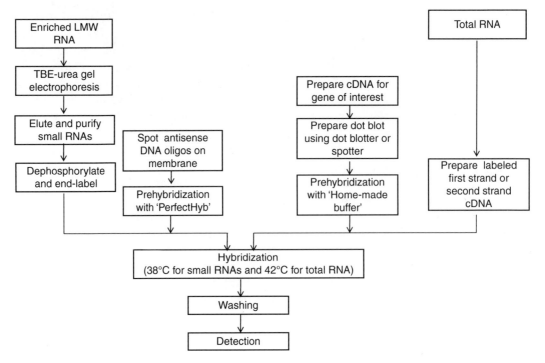

Fig. 2. Schematic representation of reverse northern blot analysis of total RNA as well as small RNAs.

sequence data generation. On the high-throughput front, expression of mRNA transcripts is easily validated by the use of microarrays. The technology for conducting microarrays of gene transcripts is fairly advanced and commercially accessible to many laboratories. However, microarray-mediated small RNA validation is largely limited to known miRNAs. Validation of the novel miRNAs and siRNAs require extensive designing of customized microarrays, technology for which is still developing. The most common medium-throughput approach to study the expression of gene transcripts, known miRNAs, novel miRNAs, and siRNAs is "reverse northern," a "reverse" format of conventional northern approach. It is a cost-effective and user-friendly method to analyze expression levels of multiple RNAs simultaneously. In this method, ESTs or cDNA fragments (in case of transcripts) or oligonucleotides (in case of small RNAs) are blotted onto a membrane by a dot-blotter or robotic spotter, which is then hybridized with radiolabeled total RNA or small RNAs as the probe, respectively (22–25). The schematic workflow of reverse northern analysis for total as well as small RNA is presented in Fig. 2.

Northern and reverse northern approaches aid in determining the abundance of steady-state level transcripts which is actually a culmination of RNA synthesis as well as its degradation. However, in some instances, it becomes necessary to study the transcription rate independently of RNA degradation/turn over, modification, and stability. Nuclear run-on transcription assay is a simple, classical approach for measuring the rate of gene transcription that provides insights into the gene regulation at transcriptional level. The basic protocol involves lysis of cells to recover the intact nuclei. The isolated nuclei contain halted transcription complexes on the DNA template mainly due to unavailability of ribonucleotides. The transcription is restarted by in vitro addition of ribonucleotides, one of which is radiolabeled. Consequently, labeled transcripts are produced, which are then hybridized to gene-specific DNA fragments that have been blotted onto a membranous support. Subsequently, labeled ribonucleotides are added in vitro to allow completion of transcription process. Nuclear run-on procedure involves multiple steps and requires good technical skills. The major step that determines the success of a run-on experiment is isolation of high quality intact nuclei, which is a challenging task for many plant systems. For complete details of this technique, please refer to ref. 26.

Though northern blotting and hybridization are relatively easy to perform in any standard molecular biology laboratory, there are inherent limitations associated with this technique. Foremost, a large quantity of RNA is required to carry out northern blotting and many a times the tissue amount is a limitation. This problem is further augmented in case of small RNA northerns as small RNAs are just a minor fraction of total RNA. Second, most of the time only one gene/small RNA can be investigated in a single experiment. Even the reverse northern technique can at best be used to study the expression of a few hundred genes/small RNAs. Therefore, the northern or its variant reverse northern are low- to medium-throughput techniques and cannot be used to interrogate the expression level changes on a genomic scale. The high-throughput techniques to quantify genomic-scale expression changes mainly rely on microarrays or new-generation sequencing technologies. Third, even though the sensitivity of detecting transcripts and small RNAs using northern blotting has substantially improved over time, this technique is less sensitive than microarrays and real-time PCR. Use of real-time PCR to detect miRNAs have been a recent development (22–31), and presents a unique challenge in designing the primers because of short length of miRNAs.

In spite of some limitations, northern analysis still remains the most widely accepted method for expression quantification studies. One major advantage of this approach is that it is the only method that visualizes the molecular size of the transcript at the same time. Besides identifying full-length transcripts, it readily reveals presence of splice variants or transcription processing machinery errors (deletions, etc.). Additionally, protocols for deprobing and reprobing the membrane (such that same blot can be used multiple times) are standard and are easily conductible. Modifications in regular northern assay produce results of unquestioned quality and reproducibility.

For example, electroblotting is an alternative and efficient method of RNA transfer from gel to membrane over traditional capillary transfer, especially in case of small RNAs. Another modification that has improved sensitivity is the use of EDC crosslinking in conjunction with LNA probes. Because of these modifications, small RNA northern can be carried out with reduced amount of RNA. The option to use LNA probes provides additional flexibility to manipulate the hybridization temperatures. Hybridization at a higher temperature reduces the background or nonspecific binding. Use of phosphorimager instead of regular X-ray films further improves the detection of the signal.

In this chapter, we have compiled detailed protocols for performing northern blot analysis for gene transcripts and small RNAs. Protocols for medium-throughput screening of transcripts and small RNAs by reverse northern are also included.

2. Materials

2.1. For Northern Analysis of Total RNA

Isolation of good quality RNA is of the paramount importance for performing northern analysis. A number of protocols are available for extraction of total RNA or high molecular weight (HMW) fraction of RNA. Mostly intact high quality RNA can be prepared using the GITC-phenol method described by Chomczynski and Sacchi (32). Detection of less abundant transcripts entails enrichment of messenger RNA, and two different affinity chromatographic methodologies are commonly used for isolation of poly(A)$^+$ mRNA. These methods either rely on oligo (dT)-cellulose chromatography or use of biotin-labeled oligo-dT primer in conjunction with streptavidin paramagnetic particles. For routine northern analysis, 10–50 μg of total RNA or 1–5 μg of enriched mRNA is employed (see Note 1).

2.1.1. Agarose–Formaldehyde Gel Electrophoresis Components

1. *Diethylpyrocarbonate (DEPC)-treated water.* Add DEPC to 1,000 mL of water to a final concentration of 0.1%. Stir overnight and then autoclave. Store either at room temperature or 4°C.

2. *Electrophoresis buffer.* 1× 3-[N-morpholino] propanesulfonic acid (MOPS) buffer. Prepare stock of 20× MOPS buffer by adding 83.6 g of MOPS in 700 mL of DEPC-water and adjust pH to 7.0 with 2 N NaOH. Add 40 mL of DEPC-treated 1 M sodium acetate (pH 5.2) and 40 mL of 0.5 M EDTA (pH 8.0). Make up the volume to 1,000 mL with DEPC-treated water (see Note 2).

3. *1.2% Agarose–formaldehyde gel.* 1× MOPS buffer, 1.11% formaldehyde, 1.2% agarose. Add 92 mL of DEPC-treated water to RNase-free beaker followed by 1.2 g of agarose. Now add

5 mL of 20× MOPS buffer and boil until agarose dissolves completely. Allow the solution to cool (up to 50°C) and then add 3 mL of 37% formaldehyde. Mix well without introducing bubbles and pour the solution in the gel caster in a fume hood. Cover the gel with cling wrap and allow it to solidify for at least 30 min at room temperature (see Note 3).

4. *High molecular weight (HMW) RNA gel loading dye.* 95% Formamide, 0.025% SDS, 0.1 μg/μL ethidium bromide, 0.5 mM EDTA, 1.25× MOPS buffer, 0.25% bromophenol blue (BPB), and 0.25% xylene cyanol (XC).

2.1.2. Blotting, Probing, and Washing Components

1. *20× Saline sodium citrate (SSC).* Add 500 mL of water to a beaker and add 175.3 g of sodium chloride and 88.2 g of tri sodium citrate. Adjust pH to 7.0 with 1 N HCl. Make up the volume to 1,000 mL with water. Filter with Whatman No. 1 and autoclave.

2. *Methylene blue staining solution.* Prepare 0.02% methylene blue in 0.3 M sodium acetate (pH 5.5) prepared in sterile nuclease-free water.

3. *Prehybridization buffer.* 50% Formamide, 5× SSC, 0.5% SDS, 5× denhardt's solution, 5% dextran sulfate. Add freshly denatured salmon sperm DNA to the final concentration of 0.2 mg/mL before use (see Note 4).

4. *Positively charged nylon membrane.* Hybond N+ (GE Healthcare, USA).

5. *Sephadex G-50 slurry.* Add 5 g of sephadex G-50 in 100 mL TE buffer and autoclave. Store at room temperature.

6. *Glass wool.* Autoclave in a glass bottle and store at room temperature.

7. *Probe preparation.* Random primer labeling kit and [α-^{32}P] dNTP are generally employed for preparing DNA probes. For preparing ribo (or RNA) probes, in vitro transcription kit is used along with [α-^{32}P]dNTP. For end labeling of oligonucleotides to be used as probes, T$_4$ polynucleotide kinase is used along with [γ-^{32}P]ATP.

8. *Wash solutions.*

   Solution I: 5× SSC, 0.1% SDS.

   Solution II: 2× SSC, 0.1% SDS.

   Solution III: 1× SSC, 0.1% SDS.

   Solution IV: 0.1× SSC, 0.5% SDS.

2.2. For Northern Analysis of Small RNA

Northern blot assay of small RNAs is somewhat challenging as compared with mRNA northern blot assay. Total RNA or enriched-small RNA can be used as the starting material for studying the expression profile of small RNAs. However, it is recommended to use enriched-small RNA as starting material in case of low abundance small RNAs expression profiling or detection. Enriched-small RNA can be obtained by precipitating total RNA with lithium chloride or PEG-NaCl (18).

2.2.1. Polyacrylamide Gel Electrophoresis Components

1. *40% Acrylamide/bisacrylamide (29 acryl:1 bis)*. Weigh 38.7 g of acrylamide and 1.3 g of bisacrylamide and transfer to a 100 mL graduated cylinder containing about 40 mL of DEPC-treated water. Make up to 100 mL with DEPC-treated water. Filter and store at 4°C in an amber glass bottle (see Note 5).
2. *Ammonium persulfate*. 25% Solution in nuclease-free water. Store at −20°C (see Note 6).
3. N,N,N,N'-tetramethyl-ethylenediamine (TEMED).
4. *15% Denaturing PAGE gel*. 7 M Urea, 15% acryl/bisacryl, 0.05% APS, 0.05% TEMED. Add 15 mL of DEPC-treated water in a graduated glass beaker. Weigh 8.4 g of urea and transfer it to the beaker. Add 2 mL of 5× TBE and warm in a water bath set at 37°C until the urea dissolves completely. Make up the volume to 20 mL with DEPC-treated water. Allow the solution to cool at room temperature and filter through Whatman No. 1 filter paper. Add 40 µL of 25% APS and mix well. Now add 10 µL of TEMED, mix well, and pour immediately. For routine small RNA northern analysis, 7.5 × 10 cm gels with 1.5 mm spacers are employed (see Note 7).
5. *PAGE gel running buffer*. 1× TBE (see Note 8).
6. *LMW gel loading dye*. 95% Formamide, 18 mM EDTA, 0.025% SDS, 0.25% xylene cyanol, and 0.25% BPB.
7. *Ethidium bromide gel staining solution*. 0.5× TBE containing 0.5 µg/mL ethidium bromide (see Note 9).

2.2.2. Electroblotting, Probing, and Washing Components

1. *Positively charged nylon membrane*. Hybond N+ (GE Healthcare, USA).
2. *Electroblotting buffer*. 0.5× TBE.
3. *Prehybridization buffer*. PerfectHyb (Sigma Chemical Company, USA).
4. T_4 polynucleotide kinase (New England Biolabs, USA).
5. *Salmon sperm DNA*. See item 3, Subheading 2.1.2.
6. *Sephadex G-25 slurry*. See item 5, Subheading 2.1.2.
7. *Glass wool*. See item 6, Subheading 2.1.2.
8. *Wash solutions and membrane*. Same as for total RNA (see items 4 and 8, Subheading 2.1.2).

2.3. Reverse Northern Dot Blot Components for Total RNA

1. *DNA denaturation solution.* 1 N NaOH.
2. *DNA neutralization solution.* 3 M Sodium acetate, pH 5.2.
3. *Dot-blotting solution.* 10× SSC.
4. Dot-blotting apparatus with vacuum pump or automated dot blotter.
5. *Nylon membrane, prehybridization solution and salmon sperm DNA.* See items 3 and 4, Subheading 2.1.2.
6. *Probe preparation.* First-strand cDNA synthesis kit, Random primer labeling kit, and [α-^{32}P]dNTP.
7. *Probe purification.* See items 5 and 6, Subheading 2.1.2.
8. *Wash solutions.* See item 8, Subheading 2.1.2.

2.4. Reverse Northern Dot Blot Components for Small RNA

1. *Dot-blotting solution and dot blotter.* See items 3 and 4, Subheading 2.3.
2. *Nylon membrane, prehybridization solution and salmon sperm DNA.* See items 3 and 4, Subheading 2.1.2.
3. *15% Polyacrylamide gel and gel electrophoresis components.* See items 4–7, Subheading 2.2.1.
4. *Gel elution buffer.* 0.3 M NaCl prepared in DEPC-treated water.
5. *Spin-X centrifuge tube filters.* Cat No. 8162 (Sigma Chemical Company, USA).
6. *Glycogen.* 20 mg/mL solution (Roche, Switzerland).
7. 3 M Sodium acetate, pH 5.2.
8. *Probe preparation.* Calf intestinal alkaline phosphatase (CIAP), T$_4$ polynucleotide kinase (T$_4$ PNK), [γ-^{32}P]ATP.
9. *Probe purification.* See items 6 and 7, Subheading 2.2.2.
10. *Prehybridization buffer, salmon sperm DNA, and wash solutions.* See items 3, 5, and 8, Subheading 2.2.2.

2.5. Deprobing Solutions

1. Wash solution I: 0.5% SDS.
2. Wash solution II: 0.1% SDS.

---

3. Methods

3.1. Northern Analysis of Total RNA

Accurate quantification of RNA is the most crucial and necessary step before starting the experiment. RNA can be quantified using spectrophotometer, nanovue/nanodrop, Quant-iT kit (Invitrogen, USA), or Bioanalyzer (Agilent Technologies, USA). We regularly use spectrophotometer and NanoVue/NanoDrop (see Note 10).

The real differences in gene transcript expression can only be observed when equal amounts of RNA are loaded into each lane of the gel. The RNA quantitation is greatly affected by its purity and therefore best quantitation are from the RNA samples devoid of contaminants, such as DNA, phenol, phenolics, oils, and carbohydrates. The spectrophotometric quantitation is not advisable for samples where RNA concentration is very low. This normally happens when RNA is isolated from small amount of tissue. Every spectrophotometer has a specific linear range of detection. Less concentrated samples are not detectable and more concentrated samples require dilution for accurate quantitation. For low concentration samples, quantitation can be performed by NanoDrop/NanoVue/Bioanalyzer/Spectrofluorometer. In addition to the RNA purity that determines equal loading, integrity (intactness) of RNA is equally critical. Degraded RNA will not present the correct expression differences between samples. To verify the intactness, the most commonly employed approach is to resolve total RNAs on a denaturing agarose gel. Intact eukaryotic RNA resolved on formaldehyde–agarose gels shows sharp bands of 28S and 18S rRNAs and a background smear of mRNAs (see Note 11). The exact ratio between 28S and 18S RNA can be easily calculated by using quantification softwares that are supplied with gel documentation systems. The user-dependent quality assessment using ratios of 28S and 18S RNAs is inconsistent and therefore an algorithm known as RNA integrity number (RIN) was developed that takes into account the entire RNA electrophoretic trace to determine the RNA integrity. This algorithm was developed by research scientists working at Agilent Technologies, USA. They also pioneered the development of an instrument, Bioanalyzer, which enables microcapillary electrophoretic RNA separation and the combination of Bioanalyzer and the RIN algorithm can be used to determine the RNA integrity and concentration in an unambiguous way (see Note 12).

3.1.1. Agarose–Formaldehyde Gel Electrophoresis

1. Prepare a 1.2% agarose–formaldehyde gel (see item 3, Subheading 2.1.1) (see Note 13).

2. Take appropriate amount of total RNA sample in a microcentrifuge tube and add equal volume of 2× HMW RNA gel loading dye. In a separate tube, set up the denaturation reaction for RNA marker. Denature the mix at 65°C for 10 min and snap chill in ice for 5 min. Briefly centrifuge the tubes to settle the tube contents.

3. Load the denatured samples and the RNA marker and start the electrophoresis (at constant 100 V) till the dyefront (the faster migrating BPB dye in the samples) has migrated two-third of the length of the gel.

4. Visualize the RNAs by placing the agarose gel on a UV transilluminator. Take a photograph along with the fluorescent ruler for the record (see Note 13).

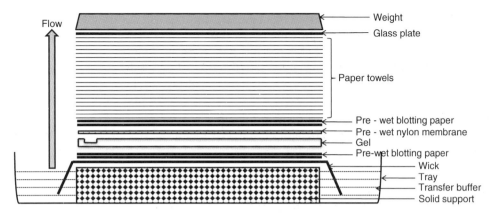

Fig. 3. Diagrammatic representation of upward capillary transfer of total RNA from denaturing agarose gel to nylon membrane. The capillary transfer is carried out at room temperature for 16 h.

3.1.2. Transfer of RNA from Agarose Gel to Nylon Membrane and Immobilization

Transferring the resolved RNAs onto the solid matrix is a crucial step in northern analysis. A number of methods such as vacuum blotting, electroblotting, semidry blotting, and capillary blotting are routinely employed for the transfer. However, capillary blotting is the most preferred method for efficient transfer of total RNAs or enriched mRNAs onto the nylon membrane as it is relatively easy and do not require any instrumentation.

1. After the RNA is sufficiently resolved, stop further electrophoresis and rinse the gel with DEPC-treated water. Incubate in 10× SSC for 10 min with gentle shaking and again rinse in DEPC-treated water. Mark the orientation of the gel.

2. Cut a piece of positively charged nylon membrane to the size of the gel and make an orientation mark so that it can be aligned with the gel orientation (see Note 14).

3. Wet the membrane with sterile water followed by its incubation in 2× SSC for at least 5 min (see Note 15).

4. Prepare a 3–4 cm stack of paper towels and four sheets of filter paper to the size of the gel. Also, cut two wicks of Whatman No. 3 filter paper depending upon the size of the container in which capillary sandwich assembly is to be set up. Wet the wick and filter paper sheets in 2× SSC.

5. Assemble the capillary transfer sandwich assembly as shown in Fig. 3. Cover the top of the stack with a thin glass plate or gel casting tray and surrounding of the gel with parafilm or cling film. Place an appropriate weight on the top of paper towels. Allow the transfer of RNA for overnight (see Note 16).

6. Disassemble the capillary transfer system and carefully mark the position of wells on the membrane with a waterproof ink pen through the gel. Briefly rinse the membrane in 2× SSC and without delay remove excess fluid. Place the membrane on a clean Whatman paper with the RNA side upward.

7. Fix the RNA on the damp membrane by placing it in a UV cross-linker (RNA side up) at $1,200 \times 100$ µJ/cm². Directly proceed with the hybridization step or wrap the membrane sandwiched between Whatman papers, with cling wrap and store at 4°C (see Note 17).

3.1.3. Prehybridization

1. Place the membrane in the hybridization bottle, with immobilized RNA side facing inwards. Remove the air bubbles between bottle and membrane with the help of blunt end forceps (see Note 18).

2. Denature 200 µL of sheared salmon sperm DNA at 95–100°C for 10 min and immediately snap cool it in ice for 5 min. Add the denatured DNA to prehybridization solution (see Note 19).

3. Prehybridize membrane for 4 h to overnight at 42°C at 30 rpm in a bottle rotisserie fitted in hybridization oven (see Note 20).

3.1.4. Probe Preparation and Purification

1. The hybridization can be performed either with labeled DNA, RNA, or oligonucleotides. The DNA probes are routinely prepared either by random priming or nick translation, whereas the in vitro transcription is the most desirable approach to prepare RNA probes. The oligonucleotides are end-labeled using T_4 polynucleotide kinase and [γ-^{32}P]ATP. Random primer labeling or nick translation may be performed using commercially available kits. Purification of probe to remove unincorporated radionucleotides is an additional step that increases the signal-to-noise ratio and is therefore recommended. The radiolabeled probe is generally purified using column chromatography with sephadex G-25 or G-50 (see Note 21). Take 1 µM of oligonucleotide probe in a 1.7-mL microcentrifuge tube. Add autoclaved milliQ water to make the volume 11 µL.

2. Add 2 µL of 10× T_4 polynucleotide kinase buffer, 5 µL of [γ-^{32}P]ATP, and 2 µL of T_4 polynucleotide kinase to the probe containing tube.

3. Incubate the tube at 37°C for 30 min.

4. After incubation, place the tube on ice. The probe is now end-labeled and ready for purification.

5. To prepare sephadex column, add small amount of glass wool into a glass Pasteur pipette and push it upto its neck to seal the opening in such a manner that only liquid can flow through it. Now add 1 mL of water into the pipette and check for the flow. It should be a slow drop by drop flow (see Note 22).

6. Add sephadex slurry in the column upto three-fourth of its length. Do not let the column dry once it is packed (see Note 23).

7. Increase the volume of probe to 100 µL with sterile nuclease-free water.

8. Add entire amount of the probe to the top of sephadex column. Wait for 30 s and then add appropriate amount of water. A continuous supply of water is maintained till the purification procedure is completed.

9. Monitor the elution of the probe from the column with a hand-held GM counter. Initially the readings in the GM counter will increase very slowly, followed by a rapid increase and then a drop. The probe collected during the rapid increase till the drop in the readings is the desired fraction and is of the highest specificity. Discard the other fractions, denature the purified probe at 80–100°C for 5 min in a screw cap tube, and snap cool in ice for 5 min (see Note 24).

3.1.5. Hybridization

1. Add purified probe to bottle containing blot and prehybridization buffer (see Note 25).
2. Rotate overnight at 42°C in an oven (see Note 26).

3.1.6. Washing and Detection

1. Discard hybridization buffer in radioactive liquid waste or store for future use (see Note 27). Rinse the blot with solution I so that residual probe sticking to the blot or membrane is washed off.
2. Wash the membrane with wash solution I at 42°C twice for 10 min each.
3. Wash the membrane with wash solution II and III at room temperature for 10 min each.
4. Monitor the blot with a hand-held GM counter. Ideally, there should be no signal in the wells or at places where there is no RNA (see Note 28).
5. In case the blot still has a lot of background counts, wash the blot with wash solution IV at room temperature initially for 5 min and if required for another 10 min (see Note 29).
6. Rinse the blot with 2× SSC and remove excess solution. Wrap the blot in cling wrap and squeeze out air bubbles along with excess liquid.
7. Expose blots to phosphorimager screen or X-ray film (see Note 30).

3.1.7. Deprobing the Membrane

1. After detection, remove cling wrap from the blots (see Note 31).
2. Place the membrane in a hybridization bottle or a tray and add stripping solution I. Rotate at 80°C for 30 min (see Note 32).
3. Pour out the used wash solution in liquid radioactive waste and again add preheated stripping solution II. Rotate at 80°C for 30 min.
4. Rinse in 2× SSC and wrap the blots. Monitor the signal with the hand-held GM counter and if signal is observed steps 2 and

3 can be repeated or extended. Cross-check the presence of signal by exposing the blots overnight to a phosphorimager screen or X-ray sheet. If any signal is detected then repeat the above steps at 80°C. The deprobed membranes are ready for probing with another probe (see Note 33).

3.2. Northern Analysis of Small RNA

For small RNA northern blot analysis, small RNAs (10–100 μg) are resolved on denaturing polyacrylamide gel and are then electroblotted onto nylon membrane which is probed with labeled oligonucleotides/LNA probes for the detection of specific miRNA/siRNA.

3.2.1. Resolution of Small RNAs by Polyacrylamide Gel Electrophoresis

1. Assemble gel casting unit and pour 15% TBE-urea polyacrylamide gel (see item 4, Subheading 2.2.1). Insert an appropriate comb and allow it to polymerize for at least 30 min (see Note 34).
2. Disassemble gel casting unit and remove comb carefully without distorting the wells.
3. Add 1× TBE buffer in gel reservoir. Flush the wells thoroughly with gel running buffer using a thick needle attached to 10-mL syringe to remove any unpolymerized acrylamide and urea (see Note 35).
4. Add 10 μL of 1× LMW gel loading dye in each well. Pre-run the gel at constant 200 V for 20 min (see Note 36).
5. While the gel is pre-running, aliquot appropriate amount of LMW RNA as well as RNA decade marker (Ambion, USA) and add equal volume of 2× LMW gel loading dye. Heat the RNA–dye mix at 65°C for 10 min. Immediately plunge in ice and centrifuge the tubes to collect the entire content just before loading.
6. Load both RNA sample and RNA size marker. Run the gel at constant 200 V for 50 min. Stop electrophoresis when the lower dye (BPB) reaches the gel front.
7. Disassemble gel plates and carefully transfer the gel to ethidium bromide staining solution. Stain for 10 min on a rocker shaker with gentle shaking followed by destaining with autoclaved water for 10 min.
8. View the gel on UV transilluminator and take a photograph of the same for record. Place a ruler alongside so as to mark the position of tRNAs, rRNAs, and dye front on the gel (see Note 37).

3.2.2. Electroblotting and Crosslinking of Small RNAs

1. Pre-wet all the sponges and four filter papers (cut to the size of gel) in electroblotting buffer. Rinse the nylon membrane in water followed by wetting with electroblotting buffer. Prepare the sandwich assembly for electroblotting as shown in Fig. 4 (see Note 38).

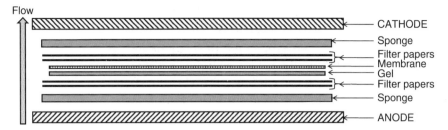

Fig. 4. Diagrammatic representation of setup for transfer of low molecular weight RNAs from denaturing polyacrylamide gel to nylon membrane by electroblotting. The transfer is carried out at 4°C and constant voltage for 16 h. Blotting time may be reduced by increasing the voltage.

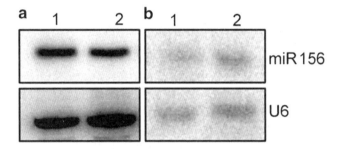

Fig. 5. Northern blot analysis of small RNAs for comparison of (**a**) PerfectHyb hybridization buffer and (**b**) home-made hybridization buffer. Low molecular weight RNAs were isolated from wheat (1) and rice (2) seedlings for resolution on 15% TBE-PAGE gel. The small RNAs were electroblotted onto positively charged nylon membrane and the membrane was probed with miR156 and U6 oligonucleotide probes.

2. Transfer the RNA by electroblotting at constant 25 V for overnight in a cold room (4°C).

3. Disassemble the unit and take out membrane very carefully (see Note 39).

4. Rinse the membrane with 0.5× TBE to remove any gel piece or particulate matter. Remove excess solution and place the membrane with RNA side up on a piece of Whatman paper followed by crosslinking in a UV cross-linker (see step 7, Subheading 3.1.2).

5. Bake at 80°C for 1 h. Keep at room temperature for 5 min, place the membrane between a folded Whatman paper. Wrap the paper with cling wrap and store at 4°C.

3.2.3. Prehybridization and Hybridization

For prehybridization and hybridization, see Subheadings 3.1.3 and 3.1.5, respectively. We routinely use PerfectHyb (Sigma Chemical Company, USA) as the prehybridization/hybridization buffer and a temperature of 38°C for detection of small RNAs. We have found that commercially available PerfectHyb solution is much better than the home-made hybridization buffer for northern analysis of small RNAs (Fig. 5).

3.2.4. Probe Preparation and Purification

For preparing oligonucleotide probes, see Subheading 3.1.4 and use sephadex G-25 for small RNA probe purification. RNA probes are usually used when the exact sequence of small RNA is not known but it is predicted that small RNA is generated from a particular region of genome. However, for detection of small RNAs, these RNA probes are hydrolyzed into smaller fragments just before use. In this case, the fragment is cloned directionally in a plasmid vector, such as pGEMT Easy, for in vitro transcription of RNA or the fragment is PCR amplified utilizing primers for bacteriophage promoter T7, SP6, or T3 polymerase. [α-^{32}P]UTP is incorporated by in vitro transcription of the desired fragment and DNA strand is removed by DNase treatment as per the manufacturer's instruction. Calculate time for carbonate hydrolysis by the following formula:

$$T(\text{time}) = (L_i - L_f)/KL_iL_f,$$

where L_i is initial length of the probe, L_f is the final/desired length of the probe, and K equals to 0.11 kb per min.

1. Add 300 μL of carbonate buffer (120 mM of Na_2CO_3 and 80 mM of $NaHCO_3$) to the 20 μL of riboprobe reaction mix, and incubate at 60°C for the calculated time.
2. Neutralize the reaction mix with 20 μL of 3 M sodium acetate (pH 5.0).
3. Add 1/10th volume of 3 M sodium acetate (pH 5.2), 2 μL of glycogen, and 2.5 volume of 100% chilled ethanol. Incubate at −20°C for 30 min.
4. Centrifuge the tube at 16,100×g for 15 min at 4°C.
5. Discard supernatant and add 500 μL of 80% ethanol to the tube and centrifuge again at same conditions for 10 min.
6. Discard supernatant completely and air dry the pellet (not more than 3–4 min). Dissolve pellet in 30 μL DEPC-treated water.
7. Heat denature the probe by placing the tube in a dry block or water bath for 5 min at 80°C.
8. Add the entire amount (30 μL) to the prehybridized membrane and proceed to hybridization at 32°C for overnight.

3.2.5. Washing and Detection

Washing of small RNA blots is carried out at 38°C. However, washing temperature can be increased upto 50°C when background and nonspecific binding is a problem. For details, please refer to Subheading 3.1.6.

3.2.6. Deprobing the Membranes

Please refer to Subheading 3.1.7. It is advisable to use fresh membranes for detecting low expressing small RNAs.

3.3. Reverse Northern for Total RNA

In this approach, DNA fragments are dot-blotted using a dot-blotter apparatus or automated spotting machine. Protocol for spotting DNA using a dot blotter aided with gentle vacuum suction is described below. These blots are then hybridized with labeled cDNA probes.

3.3.1. Preparation of Reverse Northern Dot Blot for Expression Profiling of mRNAs

1. Prepare DNA samples and denature them by adding equal volume of 1 N NaOH followed by heating at 65°C for 10 min. Snap cool on ice and neutralize by adding equal volume of DNA neutralization solution (see Note 40).

2. Cut out the nylon membrane, four sheets of Whatman filter papers of required size, and pre-wet it in 10× SSC for 10 min.

3. Assemble the apparatus and place sheets of Whatman paper followed by the membrane in the slot provided in dot-blotter apparatus. After ensuring tight seal between the paper and membrane, apply blotting solution to all the wells and check for uniform vacuum suction. Immediately apply DNA samples in the wells with gentle suction through the slots followed by applying 2× SSC to all the wells. Repeat the procedure if preparing more than one blot.

4. Disassemble the apparatus and mark the orientation of the blot and immobilize the DNA on membrane by UV crosslinking (see step 7, Subheading 3.1.2).

3.3.2. Preparation of Probes for Reverse Northern Blots

For the preparation of cDNA, total RNA or enriched poly(A) RNA is used. The RNA is reverse transcribed and the first-strand cDNA is used as a template to synthesize radiolabeled second strand by random priming. Alternatively, the first-strand cDNA can be radiolabeled by incorporating a [^{32}P]-labeled nucleotide during the reverse transcription.

1. Prepare the first-strand cDNA either by using oligo-dT primer or random primers or their combination and reverse transcriptase. We generally start with 500 ng of enriched mRNA. A number of commercial kits are available for synthesis of first-strand cDNA. Follow the manufacturer's instructions.

2. The first-strand cDNA can be radiolabeled during reverse transcription. In case the second strand is to be labeled, use first-strand cDNA as the template and synthesize radiolabeled second-strand cDNA using random primers, Klenow fragment, and [α-^{32}P]CTP.

3. Purify labeled probe through sephadex G-50 column as described in Subheading 3.1.4.

3.3.3. Prehybridization, Hybridization, Washing, and Detection

See Subheadings 3.1.3, 3.1.5, and 3.1.6.

3.4. Reverse Northern for Small RNA

Reverse northern for small RNAs are performed to determine the expression level of a number of small RNAs under a particular condition or at a developmental stage in a single experiment. The oligonucleotides complementary to small RNA sequences are spotted onto the membrane and the blots are probed with radioactively labeled low molecular weight-enriched fraction of RNA.

3.4.1. Preparation of Reverse Northern Dot Blot

For routine experiments, denatured oligonucleotides are spotted onto nylon membrane using a dot-blotter or robotic spotter (see Subheading 3.3.1).

3.4.2. Preparation of Small RNA Probe for Screening Oligonucleotide Dot Blots

Total RNAs (in the range of 20–200 µg) are resolved on polyacrylamide gel and small RNAs are purified from the gel. These purified small RNAs are then end-labeled using T_4 polynucleotide kinase enzyme in the presence of $[\gamma\text{-}^{32}P]ATP$.

1. Resolve total RNAs on a 15% TBE-urea polyacrylamide gel along with RNA size marker.
2. Following electrophoresis, stain the gel in ethidium bromide solution for 10 min, and then destain with nuclease-free water for 10 min on a rocker shaker.
3. Visualize the gel on a UV transilluminator and cut out small RNA fraction corresponding to 18–30 nt.
4. Crush the gel piece with a sterile pestle in a 2-mL microcentrifuge tube and add 400 µL of elution buffer (0.3 M NaCl). Rotate the tube for 6 h at 4°C.
5. Filter the entire contents with Spin-X centrifuge filters at $16,100 \times g$ for 4 min at 4°C. Precipitate the filtrate with equal volume of isopropanol, 1/10th volume of 3 M sodium acetate, and 1 µL of glycogen (20 mg/mL) and incubate at −20°C for overnight (see Note 41).
6. Centrifuge the tube at $16,100 \times g$ for 15 min at 4°C. Discard the supernatant and add 500 µL of 70% ethanol. Wash the pellet by inverting the tube seven to eight times. Keep at room temperature for 5 min. Centrifuge at $16,100 \times g$ for 10 min at 4°C.
7. Discard the supernatant and air dry the pellet.
8. Dissolve the pellet in 10 µL of DEPC-treated water.
9. Dephosphorylate using alkaline phosphatase (CIAP) and subsequently label with $[\gamma\text{-}^{32}P]ATP$ and T_4 polynucleotide kinase. The labeled small RNAs are then purified by passing through a sephadex G-25 column, denatured, and employed for probing the reverse northern blots.
10. For prehybridization, hybridization, washing, and detection, see Subheadings 3.2.3 and 3.2.5.

3.5. Normalization of Signal

The major application of northern analysis is to investigate expression of a particular gene in one sample with respect to another sample. To conclusively comment on the expression of the gene, it is also important to study the expression of a housekeeping gene which is included as internal control in expression profiling experiments. These internal controls are the genes whose steady-state transcript level remains unchanged under a broad range of experimental conditions or at different stages of development of an organism if equal amount of RNA is loaded onto the gel. Numerical values can be assigned to the signals obtained from the candidate gene expression and the house keeping genes using softwares supplied with densitometer/phosphorimager. The ratio between the amount of the candidate gene transcript and that of housekeeping gene is the real measure of the candidate gene expression in a particular sample. It is therefore worthwhile to identify the internal standard for any northern analysis experiment. It is advisable to perform validation of a number of internal controls in the experimental system (33, 34). Normalization of signal is crucial for interpreting the results obtained with northern analyses. For different types of northern blot assays, different controls are utilized. A brief description of the controls for total and small RNA analysis is presented below.

1. For total RNA analysis, it is recommended to probe the membrane simultaneously with the internal control as well as experimental probe if the transcript size of the candidate gene and the internal control is distinct and the signal obtained from hybridization with one gene does not overlap with the signal from the other gene. If the transcript sizes are in the similar size range, the membrane should be first probed with one gene followed by deprobing and reprobing with another gene depending upon the abundance of the transcripts. In eukaryotes, rRNAs (28S and 18S), actin, tubulin, eIF-4A are routinely employed as internal controls for normalization of RNA levels.

2. For small RNA analysis, U6, 5.8S rRNA, and 5S rRNA are used as probes for normalization. The experimental small RNA probe and standard U6 probe may be used concurrently on a single membrane. However, owing to the huge difference in the intensity of signal obtained with the two probes, it is always recommended to use these probes sequentially. A better and faster approach is to cut the membrane such that the upper portion of the blot containing tRNA and rRNA can be probed with U6 and the lower portion is probed for candidate small RNA.

3. For reverse northern analysis, the internal controls are spotted along with the experimental genes. In case of total RNA reverse northerns, DNA corresponding to actin or tubulin or eIF-4A

is spotted in few wells. In case of small RNA reverse northerns, either certain small RNAs whose expression is ubiquitously constant or DNA corresponding to U6/5S rRNA can be spotted for normalization.

4. Notes

1. It is necessary to strictly maintain RNase-free conditions throughout the experiment as RNA is extremely susceptible to RNases. The tips and tubes should be certified RNase free, the glassware and plasticware should be pretreated with DEPC and then autoclaved, and the pestle and mortars employed for homogenizing the tissue should be baked in oven at 250°C for at least 2–4 h. All the solutions and water should be treated with DEPC followed by sterilization. Handling with bare hands should be avoided during the entire isolation and handling procedure.

2. The MOPS buffer may be autoclaved and stored in amber glass bottle. However, on autoclaving, color changes to transparent yellow. The change in color does not affect the quality of buffer. Dark yellow buffer does not perform well and therefore should not be used.

3. Formaldehyde is a suspected nose, nasopharynx, and liver carcinogen. It is toxic both through inhalation and ingestion. Its use should be restricted to fume hood.

4. We prepare a 5 mg/mL stock solution of salmon sperm DNA. The DNA is added to warm water and left for dissolving on a magnetic stirrer. The complete dissolution takes 4–5 h. The DNA is then sheared by autoclaving and the solution is stored as aliquots at −20°C. Shearing can also be carried out by passing the DNA through a 23 G 0.6 mm needle for 50 times. Alternatively, DNA may be sonicated for shearing. Immediately before use, the frozen aliquots are denatured at 95°C for 5 min, followed by snap cooling in ice and then added to the prehybridization/hybridization mix.

5. Unpolymerized acrylamide is a potent neurotoxin. While weighing, wear mask and gloves and prepare the solution in fume hood. Rinse the container used for preparing solution several times after use. Acrylamide and bisacrylamide on long-term storage and exposure to light converts to acrylic and bisacrylic acid. Therefore, fresh solution should be prepared every month and stored in amber bottle or glass bottle wrapped in aluminum foil. For long-term storage, Amberlite MB 150 resin is added to the acrylamide solution and the solution is

stored at 4°C. When preparing PAGE gel solution, filter the acrylamide to separate resin from the solution.

6. It is advisable to freshly prepare APS as persulfate decomposes rapidly in solution. Old stocks of APS are known to adversely affect band quality and sharpness of the RNA bands. However, we store the solution in small aliquots at −20°C and take out a fresh aliquot every time for preparing the gel. We have used APS solution stored at −20°C for upto 2 weeks and found no significant change in the quality of gels.

7. Urea should not be warmed at very high temperatures as it undergoes decomposition.

8. A stock of 5× TBE is prepared and diluted to obtain 1× running buffer. The 5× TBE stock exhibits minor precipitation at room temperature if stored for extended durations. We prefer making just enough quantity of 5× TBE to last up to a month. Concentrated stocks (more than 5×) are usually not prepared because of greater precipitation. We have used 5× stocks showing minor precipitation without any compromise in the gel quality. However, if the precipitation is higher, usually a fresh stock is prepared.

9. Ethidium bromide is a potent mutagen. Handle ethidium bromide solution carefully. The solution should be stored in glass amber bottle. Do not reuse staining solution when performing northern analysis. Disposal of ethidium bromide containing solutions requires special precautions and wherever available the material safety personnels/department should be involved in safe disposal of these solutions.

10. UV spectrophotometer is the classical and most widely used method for RNA quantitation. Contaminants (DNA, phenol, phenolics, oils, carbohydrates, etc.) interfere with the absorbance value of RNA and therefore it is important that these contaminants are either absent or present in minimal amounts in RNA samples. DNA contamination can be removed by DNase treatment followed by phenol extraction. Phenol itself can skew the RNA absorbance readings and therefore chloroform extraction is recommended to remove the phenol contamination. For removing other types of contaminants (present in specific tissues or plants), specialized protocols for RNA extraction are employed.

11. Intact RNAs are known to exhibit 2:1 ratio of 28S:18S rRNA. Any deviation from this ratio accompanied by fuzzy rRNA bands indicates degraded RNA. However, it is noteworthy that tissue-specific variations in the rRNA ratios are known (www.ambion.com/techlib/tn/111/8.htmL) and therefore the ratio of 2 is not an absolute indicator of RNA quality. Moreover, due to storage conditions, disruption and multiple steps in

RNA extraction, it is practically very difficult to obtain the ratio of 2. Therefore, rRNA ratios of more than 1 are generally acceptable as such RNA preparations are known to yield good quality northern results.

12. Bioanalyzer is an automated, highly sensitive lab-on-a-chip electrophoresis system which is a combination of fluidics, capillary electrophoresis, and fluorescent dye. It is employed to determine RNA integrity as well as concentration. RIN obtained through Bioanalyzer is a measure of RNA quality. Intact high quality RNA has RIN close to 10. The electrophoresis analysis on this chip is useful for the samples where only limited amount of tissue is available from which very low amounts of RNA is extracted. With the availability of Pico assay chips as low as 50 ng/µL RNA can be quantified. To avoid errors in RNA quantification between samples due to inconsistencies in RNA isolation, it is recommended that all the samples to be analyzed should be quantified at the same time.

13. Eukaryotic total RNA sample will give sharp and intact bands of 28S rRNA and 18S rRNA on denaturing gel. The intensity ratio of 28S rRNA and 18S rRNA band should be 2:1 for a high quality and intact RNA sample. However, if poly(A) RNA is resolved, the gel predominantly shows smear corresponding to mRNA. The fluorescent ruler is placed to determine the migration distance of the marker which can be later utilized to determine the size of the transcript.

14. Carefully handle the membrane by the edges with blunt forceps or gloved hands to avoid any physical damage and transfer of finger prints.

15. Prewetting the nylon membrane in sterile water results in better and uniform saturation with 2× SSC.

16. Remove air bubbles between the filter papers and membrane or between gel and the membrane carefully with the help of a glass rod. Air bubbles introduce breaks in the capillary and result in nonuniform transfer of nucleic acids. Cover the sides by cling film to prevent evaporation as well as to avoid absorption of buffer directly by the paper towels. An appropriate weight is placed on the top of the stack to ensure tight connection between the layers of the sandwich for efficient transfer. Do not attempt to check the transfer as it will disturb the alignment. Downward capillary transfer is known to be faster and does not result in flattening of gels. However, in our laboratory, we employ original method of upward transfer with good results.

17. Staining of membranes with methylene blue is suggested before initiating prehybridization and hybridization process (35). Methylene blue staining is a rapid and reversible method for

visualization of RNA and marker on the gel. Sharp rRNA bands along with mRNA smear are clearly visible after destaining. Staining for short durations results in less background and also shorter destaining time. Mark the position of marker RNA, any strong bands on the membrane, and take a photograph. Complete destaining is done in sterile water or $0.1\times$ SSC and 1% SDS for 15 min at room temperature with constant shaking.

18. Hybridization bottles are preferred over trays/boxes as they require less amount of hybridization solution. Moreover, multiple blots may be probed with the same probe by placing a mesh in between the blots. Also, posthybridization washes are easier to perform in bottles as compared to trays. Nevertheless if a tray/box is used for hybridization, make sure that the lid is sealed properly to avoid evaporation of the solution. The RNA side should be facing upward and enough prehybridization/hybridizations solution to keep the blots submerged at all times is added.

19. Check that the rotisserie is rotating or the tray (if used) is also rocking continuously. Constant agitation uniformly wets the blots and prevents them from sticking to each other. Additionally, the constant rotation/agitation ensures that membrane is always wet and this prevents sticking of probe to the membrane.

20. The prehybridization/hybridization solutions may show precipitation when stored for extended durations. This precipitate is readily solubilized by prewarming the solutions at 37–42°C.

21. Many companies provide prepacked sephadex columns and probes can be readily purified using these columns and microcentrifuge. However, we have observed that probes purified by longer columns using gravity flow yield probes with higher specific activity. These columns can be readily prepared in the laboratory and are much economical than the commercially available ones.

22. Generally, 6-cm long glass Pasteur pipettes are employed for packing sephadex columns. 1-mL insulin syringes may be used in place of glass pipettes.

23. After inserting the wool, immediately add some water. Without letting the complete drain-off of water, add the sephadex slurry. Keep on adding the slurry until the desired length of sephadex is packed. Addition of slurry immediately after water helps in avoiding entrapment of air bubbles between the glass wool and the packed sephadex. Make sure that there is no break in the column. This can be achieved by not disturbing the column while packing and also not letting the column dry during and after the packing.

24. Oligonucleotide and RNA probes can be denatured at a lower temperature (70°C), whereas the double-stranded DNA probes require higher temperatures (90°C). The screw cap tubes are advantageous as at higher temperatures the caps of the microcentrifuge tubes pop up and can cause spillage. Additionally, it is advisable to use filter tips whenever working with radioisotopes as this greatly reduces the pipette contaminations by aerosols.

25. While adding the denatured probe to the hybridization bottle, the probe is premixed in minimal amount of prehybridization buffer and then this mixture is added to the bottle. Avoid adding the purified probe directly to the bottle or onto the blot.

26. If nonspecific bands or high background along with the specific signal is observed, hybridization can be carried out at higher temperature of up to 50°C.

27. The hybridization buffer containing probe may be reused two to three times or more depending upon the half-life of the radioisotope. In that case, store the hybridization solution in a tube at −80°C. The tube containing probe should be placed in radiation-safe boxes. Before use, thaw the solution at 37°C and add it to the bottle containing blot.

28. If the background noise is high, washings with solution II and solution III can be carried out at 37–42°C before proceeding to washing with solution IV.

29. High temperature washings with solution IV can also be carried out; however, extreme care should be observed in doing so as even the specific signal can be washed off at such a high stringency.

30. Phosphorimager-based detection has advantage over X-ray film based detection with respect to increased sensitivity and short exposure time. It also gives a digital image of the blot, which may be readily utilized for presentations. More importantly, the software available with the phosphorimager is useful for accurate quantification of the signals obtained by northerns and subsequently for relative quantifications. Additionally, use of phosphorimaging is an environment friendly option.

31. The cling wrap helps in keeping the membrane moist at all times for a few days. Extended storage of the membrane in the cling wrap results in dry membranes, from which the probe removal can be problematic. In case storage of membranes for longer durations is required, store the deprobed membrane in 2× SSC solution at 4°C.

32. Preheated deprobing solution at 80°C is useful. Perform the stripping protocol preferably in a tube or a tray with lid so as to avoid spillage resulting in radioactive contamination.

33. The stripped membrane can be used for two to three more probings. It is recommended to use the probes in order of expected low to high abundance signals.

34. Avoid any air bubbles while inserting the comb. Air bubbles deform the wells and thereby affect band quality. Add a little excess gel solution after inserting the comb to compensate for gel shrinkage upon polymerization.

35. Make sure the wells are clean and no loose pieces of gel or any urea crystals are present in the wells. Well-formed and clean wells are the key parameters for obtaining good and sharp band quality.

36. Prerunning the gel not only provides denaturing conditions for good resolution of small RNAs but also gives an idea about the well quality. When loading RNA samples and the size marker, use only those wells which look good during prerunning.

37. Three types of low molecular weight RNAs are clearly visible as bands on the PAGE gels: 5.8S rRNA (~150 nt), 5S rRNA (~120 nt), and tRNAs (~80–90 nt). Small RNAs are visible as faint smear below the tRNAs on PAGE gels.

38. Remove bubbles from Whatman/gel/membrane/Whatman sandwich by rolling a clean glass tube over it. Once assembled, do not disturb the sandwich assembly.

39. View the gel on a UV transilluminator to confirm efficient transfer. The gel may be stained briefly to check for any traces of RNAs.

40. For reverse northerns, plasmid DNA containing gene of interest may be linearized using appropriate restriction enzymes or PCR amplified fragment may be employed. When using linearized plasmid DNA, make sure to include linearized empty plasmid vector as a background control. Denaturation of DNA by heating at 80–90°C for 5 min followed by quick cooling on ice is another method. BPB dye is added in DNA denaturation solution so as to visualize the DNA samples on membrane while spotting. Generally DNA is diluted to a concentration of 50 ng/μL with TE buffer. Finally, approximately 100–150 ng of DNA may be spotted on the membrane.

41. Glycogen is used as a coprecipitant during ethanol precipitation of low amount of nucleic acids.

Acknowledgments

Research work in the laboratories of S.K.-A. and M.A. is supported by grants from Department of Biotechnology (DBT), India and University of Delhi. R.P. and A.R.B. are thankful to Council of Scientific and Industrial Research (CSIR), India and DBT, India for the student fellowship, respectively.

References

1. Sunkar R, Zhu JK (2004) Novel and stress-regulated microRNAs and other small RNAs from Arabidopsis. Plant Cell 16:2001–2019
2. Hsieh LC et al (2009) Uncovering small RNA-mediated responses to phosphate deficiency in Arabidopsis by deep sequencing. Plant Physiol 151:2120–2132
3. Zhu QH et al (2008) A diverse set of microRNAs and microRNA-like small RNAs in developing rice grains. Genome Res 18:1456–1465
4. Bo W et al (2009) Novel microRNAs uncovered by deep sequencing of small RNA transcriptomes in bread wheat (*Triticum aestivum* L.) and *Brachypodium distachyon* (L.) Beauv. Funct Integr Genomics 9:499–511
5. Lee H et al (2010) Genetic framework for flowering-time regulation by ambient temperature-responsive miRNAs in Arabidopsis. Nucleic Acids Res 38:3081–3093
6. Moldovan D et al (2009) Hypoxia-responsive microRNAs and trans-acting small interfering RNAs in Arabidopsis. J Exp Bot 61:165–177
7. Marin E et al (2010) miR390, Arabidopsis TAS3 tasiRNAs, and their AUXIN RESPONSE FACTOR targets define an autoregulatory network quantitatively regulating lateral root growth. Plant Cell 22:1104–1117
8. Gao P et al (2010) Over-expression of osa-MIR396c decreases salt and alkali stress tolerance. Planta 231:991–1001
9. Priyanka B et al (2010) Characterization of expressed sequence tags (ESTs) of pigeon pea (*Cajanus cajan* L.) and functional validation of selected genes for abiotic stress tolerance in *Arabidopsis thaliana*. Mol Genet Genomics 283:273–287
10. Wu T et al (2010) Transcriptome profile analysis of floral sex determination in cucumber. J Plant Physiol 167:905–913
11. Zang Q et al (2010) Isolation and characterization of a gene encoding a polyethylene glycol-induced cysteine protease in common wheat. J Biosci 35:379–388
12. Katiyar-Agrawal S et al (2006) A pathogen-inducible endogenous siRNA in plant immunity. Proc Natl Acad Sci USA 103:18002–18007
13. Katiyar-Agarwal S, Gao S, Vivian-Smith A (2007) A novel class of bacteria-induced small RNAs in Arabidopsis. Genes Dev 21:3123–3134
14. Katiyar-Agarwal S, Jin H (2007) Discovery of pathogen-regulated small RNAs in plants. In: Methods in enzymology, 427:215–227. Elsevier, Amsterdam
15. Pall GS et al (2007) Carbodiimide-mediated cross-linking of RNA to nylon membranes improves the detection of siRNA, miRNA and piRNA by northern blot. Nucleic Acids Res 35:e60
16. Beckmann BM et al (2010) Northern blot detection of endogenous small RNAs (~14 nt) in bacterial total RNA extracts. Nucleic Acids Res 38:e147
17. Buhtz A et al (2010) Phloem small RNAs, nutrient stress responses, and systemic mobility. BMC Plant Biol 10:64
18. Xin M et al (2010) Diverse set of microRNAs are responsive to powdery mildew infection and heat stress in wheat (*Triticum aestivum* L.). BMC Plant Biol 10:123
19. Lu et al (2008) Genome-wide analysis for discovery of rice microRNAs reveals natural antisense microRNAs (nat-miRNAs). Proc Natl Acad Sci USA 105:4951–4956
20. Kim SW et al (2010) A sensitive non-radioactive northern blot method to detect small RNAs. Nucleic Acids Res 38:e98
21. Petersen M, Wengel J (2003) LNA: a versatile tool for therapeutics and genomics. Trends Biotechnol 21:74–81
22. Zou X et al (2010) Identification of transcriptome induced in roots of maize seedlings at the late stage of waterlogging. BMC Plant Biol 10:189
23. Jin H (2010) Screening of genes induced by salt stress from Alfalfa. Mol Biol Rep 37:745–753
24. Valdes-Lopez O (2010) MicroRNA expression profile in common bean (Phaseolus vulgaris) under nutrient deficiency stresses and manganese toxicity. New Phytol 187:805–818
25. Tang X et al (2007) A simple array platform for microRNA analysis and its application in mouse tissues. RNA 13:1803–1822
26. Meng L, Lemaux PG (2003) A simple and rapid method for nuclear run-on transcription assays in plants. Plant Mol Biol Rep 21:65–71
27. Schmittgen TD et al (2004) A high-throughput method to monitor the expression of microRNA precursors. Nucleic Acids Res 32:e43
28. Shi R, Chiang VL (2005) Facile means for quantifying microRNA expression by real-time PCR. Biotechniques 39:519–525
29. Fu HG et al (2006) A novel method to monitor the expression of microRNAs. Mol Biotechnol 32:197–204
30. Chen C et al (2005) Real-time quantification of microRNAs by stem–loop RT–PCR. Nucleic Acids Res 33:e179
31. Varkonyi-Gasic E et al (2007) Protocol: a highly sensitive RT-PCR method for detection and quantification of microRNAs. Plant Methods 3:12

32. Chomczynski P, Sacchi N (1987) Single-step method of RNA isolation by acid guanidinium thiocyanate-phenol-chloroform extraction. Anal Biochem 162:156–159
33. Jain M (2009) Genome-wide identification of novel internal control genes for normalization of gene expression during various stages of development in rice. Plant Sci 176:702–706
34. Garg R et al (2010) Validation of internal control genes for quantitative gene expression studies in chickpea (*Cicer arietinum* L.). Biochem Biophys Res Commun 396: 283–288
35. Herrin DL, Schmidt GW (1998) Rapid, reversible staining of northern blots prior to hybridization. Biotechniques 6:196–198

Chapter 3

Construction of RNA-Seq Libraries from Large and Microscopic Tissues for the Illumina Sequencing Platform

Hagop S. Atamian and Isgouhi Kaloshian

Abstract

Second-generation DNA sequencing platforms have emerged as powerful tools in biological research. Their high sequence output at lower cost and minimal input DNA requirement render them suitable for broad applications ranging from gene expression studies to personalized clinical diagnostics. Here, we describe the preparation of cDNA libraries, from both whole aphid insects and their microscopic salivary gland tissues, suitable for high-throughput DNA sequencing on the Illumina platform.

Key words: Second-generation sequencing, RNA from minute amount of tissue, Illumina, cDNA synthesis

1. Introduction

Classical DNA sequencing, based on the Sanger method, improved the diagnosis of diseases and contributed to our understanding of the various biological processes, identification of new drug targets, and classification of organisms. The launch of new sequencing technology referred to as "second-generation" sequencing made possible the incorporation of sequencing in even more studies aimed at finding answers to numerous biological questions, something which would not have been economically and rationally practical before (1). Second-generation sequencing technology is currently commercially available on five major platforms: AbI3730xl Genome Analyzer, Roche (454), Illumina Genome Analyzer, ABI-SOLiD, and HeliScope (2). Regardless of the platform, one of the bottlenecks for next-generation sequencing is the amount of time and resources required for template and library preparation.

Here, we describe the stepwise preparation of a cDNA library (also referred to as an mRNA-Seq library) suitable for sequencing with the Illumina Genome Analyzer platform. Using this protocol, we prepared high-quality mRNA-Seq libraries from whole aphids as well as microdissected aphid salivary gland tissues.

2. Materials

Prepare all solutions using diethyl pyrocarbonate (DEPC)-treated water to prevent RNA degradation by RNase enzymes. To further minimize RNA degradation, keep samples on ice at all times, unless indicated otherwise. Wear gloves and use sterile techniques when working with RNA. First-time RNA users are encouraged to read "General Remarks on Handling RNA." All glassware and plasticware should be RNase free. All centrifugation steps are performed at 4°C, unless indicated otherwise.

1. DEPC-treated water: Add 1 ml of 0.1% DEPC to 1,000 ml distilled water (see Note 1). Mix well on a magnetic stirrer overnight at room temperature until completely dissolved. Autoclave and let it cool to room temperature before use.
2. TRIzol (Invitrogen).
3. Chloroform.
4. 100% Isopropanol.
5. Nucleic acid carrier (linear polyacrylamide).
6. 70% Ethanol (EtOH), diluted with RNase-free DEPC-treated water.
7. RNeasy kit (containing RLT, RW1, and RPE Buffers; QIAGEN).
8. QIAshredder (QIAGEN).
9. DNaseI enzyme and DNaseI buffer.
10. Phenol.
11. 3 M sodium acetate (NaOAc), pH 5.2.
12. 100% EtOH.
13. mRNA-Seq Sample Prep Kit (containing Sera-mag oligo(dT) beads, Bead Binding Buffer, Bead Washing Buffer: 10 mM Tris–HCl, 5× Fragmentation Buffer, Fragmentation Stop Solution, Glycogen, Random Primers, 5× First-Strand Buffer, 25 mM dNTP, RNaseOUT, GEX Second Strand Buffer, RNaseH, DNA Pol I, 10× End Repair Buffer, T4 DNA Polymerase, Klenow DNA Polymerase, T4 Polynucleotide Kinase, 10× A-Tailing Buffer, 1 mM dATP, Klenow Exo-enzyme, 2× Rapid T4 DNA Ligase Buffer, 1.5 µM PE Adapter Oligo Mix, T4 DNA Ligase, 5× Phusion Buffer, 25 µM PCR

Primer PE 1.0, 25 µM PCR Primer PE 2.0, Phusion DNA Polymerase; Illumina).

14. 100 mM DTT.
15. SuperScript III Reverse Transcriptase (Invitrogen).
16. Elution buffer: 10 mM Tris–HCl, pH 8.5.
17. PCR Purification Kit (QIAGEN).
18. MinElute PCR Purification Kit (QIAGEN).
19. Agarose: Regular and high resolution for separation of small nucleic acids.
20. 1× Tris acetate–EDTA (TAE) buffer: 40 mM Tris acetate and 1 mM EDTA.
21. Loading dye.
22. 100 bp DNA ladder.
23. Gel Extraction Kit (QIAGEN).
24. GeneCatcher Disposable Gel Excision Kit (The Gel Company).

3. Methods

3.1. Total RNA Extraction from Aphid Salivary Gland Tissue

1. Collect microdissected salivary gland tissue in a 1.5-ml Eppendorf tube containing 100 µl of TRIzol reagent on ice.
2. Homogenize the tissue thoroughly with a small pestle for 2 min at room temperature (see Note 2). Rinse the pestle with an additional 100 µl of TRIzol reagent into the tube to remove excess tissue and proceed quickly to the next step. The tube can be frozen immediately in liquid nitrogen and stored at −80°C until use.
3. Vortex the tube thoroughly and incubate for 5 min at room temperature to allow for the complete dissociation of nucleo-protein complexes.
4. Add 40 µl of chloroform (1/5th the volume of TRIzol) and vortex vigorously for 20 s.
5. Incubate the tube at room temperature for 5 min. Centrifuge at $12,000 \times g$ for 15 min.
6. Carefully transfer the supernatant to a fresh tube (see Note 3). Add equal volume of isopropanol and 25 µg of a carrier for nucleic acids, such as linear polyacrylamide, to maximize recovery.
7. Mix by inverting (DO NOT VORTEX), spin briefly (1–2 s), and store the tube at −20°C overnight for RNA precipitation.
8. Centrifuge the tube at $12,000 \times g$ for 10 min. Carefully pipette out the liquid (see Note 4).

9. Wash the pellet with 1 ml of 70% EtOH, invert the tube three to four times, and centrifuge at $7,500 \times g$ for 10 min.

10. Pipette out the EtOH without disturbing the pellet (see Note 5). Let the tube dry on ice for 10 min and resuspend the pellet in 10–15 µl of RNase-free DEPC-treated water for 30 min.

3.2. Total RNA Extraction from Whole Aphids

1. For RNA isolation from large amount of tissue such as whole aphids, the RNeasy kit is used and steps 2–11 are carried out at room temperature, including centrifugation.

2. Grind 40 adult aphids to a fine powder in a 1.5-ml Eppendorf tube, submerged in liquid nitrogen, using a small pestle attached to an electric drill (see Note 2).

3. Transfer 100 mg of the powder to a 2-ml Eppendorf tube containing 1 ml of the RLT Buffer.

4. Homogenize the mixture by passing ten times through an 18-gauge needle. Add an additional 1 ml of the RLT Buffer to the homogenate.

5. Pass the homogenate through the QIAshredder by centrifuging at $4,000 \times g$ for 2 min. Transfer the supernatant to a clean tube.

6. Add one volume of 70% EtOH to the homogenized lysate and mix immediately by shaking vigorously.

7. Immediately apply the sample to an RNeasy midi column. Centrifuge at $4,000 \times g$ for 5 min and discard the flow through.

8. Add 4 ml of the RW1 Buffer to the RNeasy column and centrifuge at $4,000 \times g$ for 5 min to wash the column. Discard the flow through.

9. Add 2.5 ml of the RPE Buffer to the RNeasy column and centrifuge at $4,000 \times g$ for 2 min. Discard the flow through.

10. Add another 2.5 ml of the RPE Buffer to the RNeasy column and centrifuge at $4,000 \times g$ for 5 min. Discard the flow through. This helps dry the RNeasy silica-gel membrane.

11. Add 150–250 µl of RNase-free DEPC-treated water to elute the RNA. Incubate for 1 min and centrifuge at $4,000 \times g$ for 3 min.

3.3. DNase Treatment

1. Add 20 µl of DNaseI Buffer and 4 U of DNaseI enzyme to 20 µg of RNA and adjust the final volume to 200 µl with RNase-free DEPC-treated water.

2. Incubate the tube at 37°C for 30 min.

3. Add 300 µl of RNase-free DEPC-treated water to bring the final volume to 500 µl. Add 1/2 volume of phenol and 1/2 volume of chloroform. Vortex vigorously.

4. Centrifuge at 12,000 × g for 15 min.

5. Transfer the supernatant to a fresh tube. Add 1/10 volume of 3 M NaOAc, two volumes of 100% EtOH, and 25 μg of the nucleic acid carrier.

6. Incubate the tube at −20°C for at least 2 h.

7. Centrifuge at 18,000 × g for 30 min.

8. Carefully pipette out the liquid (see Note 4).

9. Wash the pellet with 1 ml of 70% EtOH and invert the tube three to four times. Incubate the tube on ice for 5 min before centrifugation to dissolve all the salts.

10. Centrifuge at 12,000 × g for 5 min.

11. Pipette out the EtOH without disturbing the pellet (see Note 5). Air dry the pellet on ice for 10 min and resuspend in 10–15 μl of RNase-free DEPC-treated water for 30 min on ice.

12. Amplify a housekeeping gene with PCR using 1 μl of the DNaseI-treated RNA as template to ascertain that there is no DNA contamination (see Note 6).

13. Check the quality and the concentration of the DNase-free RNA using instruments that measure low quantities of nucleic acids, such as an Agilent 2100 Bioanalyzer (Agilent Technologies) (see Note 7). Store the RNA at −80°C.

3.4. mRNA Purification from Total RNA

The reagents used in the steps below are provided either by the Ilumina's mRNA-Seq Sample Prep Kit or listed in materials. Illumina recommends to start with 1–10 μg of total RNA. We have constructed libraries from 100 ng and 4 or 7 μg of total RNA as starting material. The recommended quantities of the reagents are for 1–10 μg of total RNA. Use half the amount of the recommended reagents for libraries with starting material of <1 μg RNA.

1. Thaw the DNaseI-treated RNA sample on ice for 10 min. Adjust the volume of the total RNA with RNase-free DEPC-treated water to 50 μl in a 1.5-ml, RNase-free, low-retention Eppendorf tube. Keep the RNA sample on ice.

2. Heat the RNA sample at 65°C for 5 min to disrupt the RNA secondary structures and place the tube immediately back on ice.

3. Aliquot 50 μl of the Bead Binding Buffer into a fresh 1.5-ml, RNase-free, low-retention Eppendorf tube and keep until step 10.

4. Aliquot 50 μl of the Sera-mag oligo(dT) beads into a 1.5-ml, RNase-free, low-retention tube (see Notes 8–10). Put the tube on the magnetic stand (Dynal MPC®) to capture the beads and pipette out the solution.

5. Wash the beads twice with 100 μl of the Bead Binding Buffer and pipette out the solution.

6. Resuspend the beads in 50 μl of the Bead Binding Buffer and add to it the 50 μl RNA sample from step 3.

7. Rotate the tube at room temperature for 5 min (see Note 11) and put on the magnetic stand to capture the mRNA bound to the beads. Pipette out the solution.

8. Wash the beads twice with 200 μl of the Bead Washing Buffer and pipette out the solution.

9. Add 50 μl of the 10 mM Tris–HCl to the beads and heat at 80°C for 2 min to elute the mRNA from the beads.

10. Immediately put the tube on the magnetic stand and transfer the solution (containing the mRNA) to the tube with the Bead Binding Buffer prepared in step 3.

11. Heat the sample at 65°C for 5 min to disrupt the RNA secondary structures and place on ice. In the meantime, add 200 μl of the Bead Washing Buffer to the beads to prevent drying.

12. After placing the sample on ice, wash the used beads twice with 200 μl of the Bead Washing Buffer and pipette out the solution.

13. Reload the eluted 100 μl mRNA sample from step 12 to the beads, and rotate the tube at room temperature for 5 min to recapture the mRNA on the beads (see Note 12). Put the tube on the magnetic stand and pipette out the solution.

14. Wash the beads twice with 200 μl of the Bead Washing Buffer and pipette out the solution.

15. Add 17 μl of the 10 mM Tris–HCl to the beads and heat at 80°C for 2 min to elute the mRNA from the beads.

16. Immediately put the tube on the magnetic stand and transfer 16 μl of the solution containing the mRNA to a fresh PCR tube.

17. The mRNA can be stored at −80°C for later use.

3.5. mRNA Fragmentation

1. Add 16 μl of the mRNA and 4 μl of the 5× Fragmentation Buffer to a PCR tube (see Note 13).

2. Incubate the tube in a thermal cycler at 94°C for exactly 5 min. The thermal cycler should be at 94°C before putting the tube in to avoid ramping time.

3. Immediately add 2 μl of the Fragmentation Stop Solution and place the tube on ice.

4. Transfer the solution to a 1.5-ml Eppendorf tube and add 2 μl of 3 M NaOAc (pH 5.2), 40 μg of the glycogen, and 60 μl of 100% EtOH. Incubate the tube at −80°C for 30 min.

5. Centrifuge the tube at 20,200 ×g for 25 min at 4°C.
6. Carefully pipette out the supernatant without disturbing the mRNA pellet (see Note 14).
7. Wash the pellet with 300 μl of 70% EtOH.
8. Centrifuge the pellet and carefully pipette out the 70% EtOH as thoroughly as possible without disturbing the mRNA pellet.
9. Air dry the pellet for 10 min on ice.
10. Resuspend the mRNA in 11.1 μl of RNase-free DEPC-treated water.

3.6. First-Strand cDNA Synthesis

1. Add 3 μg of random primers to the mRNA tube. The total volume should be 12.1 μl.
2. Incubate the sample in a thermal cycler at 65°C for 5 min and place on ice for at least 1 min.
3. In a separate tube, mix the following reagents in the order listed: 5× First-strand buffer (4 μl), 100 mM DTT (2 μl), 25 mM dNTP mix (0.4 μl), and RNaseOUT (20 U). The total volume should be 6.9 μl; however, prepare 10% extra reagent mix if you are preparing multiple samples.
4. Add the 6.9 μl mixture to the PCR tube from step 2 and mix well.
5. Heat the sample at 25°C in a thermal cycler for 2 min.
6. Add 200 U of reverse transcriptase enzyme (SuperScript®III) to the sample and incubate the sample in a thermal cycler using the following program: 25°C for 5 min, 50°C for 50 min, 70°C for 15 min, and hold at 4°C.
7. Place the tube on ice. If needed, the samples can be stored at −20°C until the next day.

3.7. Second-Strand cDNA Synthesis

1. Add to the first-strand cDNA synthesis mix: RNase-free DEPC-treated water (62.8 μl), GEX Second Strand Buffer (10 μl), and 25 mM dNTP mix (1.2 μl).
2. Mix well and incubate on ice for 5 min or until well chilled.
3. Add the following reagents: RNase H (2 U) and DNA Pol I (50 U).
4. Mix well and incubate at 16°C in a thermal cycler for 2.5 h.
5. Use the PCR purification kit to purify the sample and elute in 50 μl of elution buffer (see Note 15).

3.8. Perform End Repair

1. Prepare the following 100 μl reaction mix in a 1.5-ml Eppendorf tube: Eluted DNA (50 μl), water (27.4 μl), 10× End Repair Buffer (10 μl), 25 mM dNTP mix (1.6 μl), T4 DNA Polymerase

(15 U), Klenow DNA Polymerase (5 U), and T4 Polynucleotide Kinase (50 U).

2. Incubate the sample at 20°C for 30 min.

3. Use a PCR purification kit to purify the sample and elute in 32 μl of elution buffer.

3.9. Adding "A" Bases to the 3′ End of the DNA Fragments

1. Prepare the following 50 μl reaction mix in a 1.5-ml Eppendorf tube: Eluted DNA (32 μl), 10× A-Tailing Buffer (5 μl), 1 mM dATP (10 μl), and Klenow Exo- enzyme (15 U).

2. Incubate the sample at 37°C for 30 min.

3. Use the MinElute PCR Purification Kit to purify the sample and elute in 23 μl of elution buffer.

3.10. Adapter Ligation

1. Prepare the following 50 μl reaction mix in a 1.5-ml Eppendorf tube: Eluted DNA (23 μl), 2× Rapid T4 DNA Ligase Buffer (25 μl), 15 μM PE Adapter Oligo Mix (1 μl), and T4 DNA Ligase (600 U).

2. Incubate the sample at room temperature for 15 min.

3. Use the MinElute PCR Purification Kit to purify the sample and elute in 10 μl of elution buffer.

3.11. Purifying cDNA Templates

1. Prepare a 50 ml 2% high-resolution agarose gel in 1× TAE.

2. Load 10 μl of the DNA eluate from the ligation step mixed with a loading dye in a center lane flanked by 100 bp DNA ladder on both sides (see Note 16).

3. Run the gel at 120 V for 60 min at 4°C (see Note 17).

4. Excise the desired region of the gel with the GeneCatcher and recover the gel slice by centrifuging the GeneCatcher into an Eppendorf tube.

5. Use a gel extraction kit to purify the sample and elute in 30 μl elution buffer.

3.12. Enrichment of the Purified cDNA Templates by PCR

1. Prepare the following PCR mix in a PCR tube: 5× Phusion Buffer (10 μl), 25 μM PCR Primer PE 1.0 (1 μl), 25 μM PCR Primer PE 2.0 (1 μl), 25 mM dNTP mix (0.5 μl), Phusion DNA Polymerase (1 U), and water to bring the total volume to 20 μl.

2. Add to the PCR mix 30 μl of the purified ligation mix (from Subheading 3.11, step 5).

3. Amplify using the following PCR protocol: 30 s at 98°C, 15 cycles of 10 s at 98°C, 30 s at 65°C, 30 s at 72°C, and 5 min at 72°C, and hold at 4°C.

4. Use a PCR purification kit to purify the sample and elute in 30 μl of elution buffer.

3.13. Validating the Library

1. Load 2 μl of the purified PCR sample on a 2% agarose gel and check the size and the purity of the final product. The final product should be visible as a distinct band at the expected size.
2. Clone 1 μl of the product and sequence several clones using Sanger sequencing to confirm the nature of the amplified products (see Note 18).

4. Notes

1. Work with DEPC under a fume hood as it is a carcinogen. The DEPC becomes harmless and is completely inactive after autoclaving.
2. It is necessary for the small pestle to exactly fit the base of a 1.5-ml Eppendorf tube. Therefore, try a few 1.5-ml Eppendorf tubes and choose the best fitting. While grinding, turn the pestle left and right while pressing against the bottom of the tube. The pestle should not turn easily and should make a sharp noise as it rubs against the wall of the tube. This ensures complete homogenization and maximum RNA yield.
3. Leave some supernatant behind to make sure that you do not carry any phenol to the next step. Trace amounts of phenol inhibit the activity of the reverse transcriptase enzyme.
4. Linear polyacrylamide is an efficient neutral carrier but does not adhere tightly to the bottom of the Eppendorf tube. Remove the supernatant by careful pipetting in order not to discard the pellet.
5. Remove the EtOH with a 1-ml pipette tip starting from the top until around the 50-μl mark. Then, use a 10-μl pipette tip to remove the rest. The key is not to leave any EtOH in the tube and at the same time not to disturb the pellet. EtOH interferes with subsequent reactions.
6. The most important step in preparing a high-quality library is to start with a high-quality RNA. Do not skip any quality control step (steps 12 and 13) to save time. Sometimes, a single DNaseI treatment is not sufficient to eliminate DNA contamination and a second DNaseI treatment is needed.
7. The RNA yields obtained from a minute quantity of tissue, such as the salivary glands, are out of detection range of UV spectrophotometers, including NanoDrop (Thermo Scientific). Correct quantification can be obtained using an Agilent 2100 Bioanalyzer (Agilent Technologies). For RNA extraction from minute quantity of tissue, use a pico range Bioanalyzer chip.

8. The kit recommends 15 μl of the Sera-mag oligo(dT) beads. However, we found 15 μl not to be sufficient to efficiently capture the mRNA from 7 μg total RNA.
9. Do not put the beads on ice as the buffers used with the beads precipitate. Therefore, steps 3–16, Subheading 3.4, should be carried out at room temperature.
10. When preparing libraries from multiple samples, process one sample at a time to ensure that the beads are not dried during the process. Time exactly 1 min for each step using a timer.
11. Hold the tube at a 15° angle and role with fingertips. Do not invert the tube as this disperses the beads throughout the tube.
12. This recapturing step ensures that captured nucleic acids are mRNA with a poly-A tail.
13. The Illumina kit recommends to use 100 ng mRNA, although we have successfully prepared a library from 5 ng mRNA.
14. At this point, no pellet may be visible, especially when starting with low mRNA concentrations. Always place the tubes in the centrifuge with the hinges facing outward. When pipetting out solutions, keep the tip away from the hinge side of the tube, where the RNA pellet should be located.
15. Following Illumina's recommendation, we have used PCR and gel purification kits from QIAGEN.
16. This step is very critical and should be handled carefully. For an unknown reason, the sample tends to float out of the well while loading. To make sure that the sample stays in the well, place the pipette tip close to the bottom of the well and release the sample very slowly. Using a DNA marker on both sides of the sample helps locate the precise gel area to be excised, as no nucleic acid smear will be visible in the sample lane.
17. Running the gel at 4°C makes the gel firmer and facilitates the excision of an accurate size band.
18. Sequence at least six to ten clones. In our experience, sequencing from high-quality libraries shows that the sequenced clones are different and not ribosomal.

Acknowledgment

This work is supported by USDA-NIFA grant 2009-05242 to IK.

References

1. Metzker ML (2010) Sequencing technologies – the next generation. Nat Rev Genet 11:31–46

2. Morozova O, Hirst M, Marra MA (2009) Applications of new sequencing technologies for transcriptome analysis. Annu Rev Genomics Hum Genet 10:135–151

Chapter 4

Strand-Specific RNA-seq Applied to Malaria Samples

Nadia Ponts, Duk-Won D. Chung, and Karine G. Le Roch

Abstract

Over the past few years only, next-generation sequencing technologies became accessible and many applications were rapidly derived, such as the development of RNA-seq, a technique that uses deep sequencing to profile whole transcriptomes. RNA-seq has the power to discover new transcripts and splicing variants, single-nucleotide variations, fusion genes, and mRNA level-based expression profiles. Preparing RNA-seq libraries can be delicate and usually obligates buying expensive kits that require large amounts of stating materials. The method presented here is flexible and cost-effective. Using this method, we prepared high-quality strand-specific RNA-seq libraries from RNA extracted from the human malaria parasite *Plasmodium falciparum*. The libraries are compatible with Illumina®'s sequencers Genome Analyzer and Hi-Seq. The method can, however, be easily adapted to other platforms.

Key words: Strand-specific RNA-seq, High-throughput sequencing, Malaria, *Plasmodium falciparum*, Splicing variant discovery, Transcript discovery

1. Introduction

The advent of high-throughput sequencing technologies marked the beginning of a new era for whole genome analysis. The cost for sequencing a genome dropped considerably over the past 5 years, a revolution for labs focusing on genome mining. Applications were rapidly derived, applying deep sequencing to various "omics," such as the development of RNA-seq to analyze whole transcriptomes. Where microarray-based techniques proved to be powerful tools in exploring gene expression profiles (1, 2), RNA-seq has the power to establish expression profiles in a more quantitative manner and to discover new transcripts and splicing variants (3, 4), single-nucleotide variations, and fusion genes at the single-base resolution. The dark side of the application is the considerable amount of complex computations that accompany RNA-seq. In addition, where preparing gDNA libraries is robust and affordable with a wide range of reagent options on the market, preparing RNA-seq libraries is more expensive and can be more challenging.

Here, we present a method that has the double advantage to use reagents originally designed for genomic DNA library preparation and to ultimately provide strand-specific information that simplifies downstream analysis. The described method is used to prepare, in a flexible and cost-effective manner, high-quality libraries from small amounts of RNA extracted from the human malaria parasite *Plasmodium falciparum* to be sequenced on Illumina®'s sequencers Genome Analyzer and Hi-Seq. It can, however, be adapted to a wide array of other organisms and platforms.

2. Materials

All materials and reagents must be of molecular biology grade and nuclease free. All solutions must be freshly prepared before each experiment. Lab benches and pipettes must be clean. The regular use of cleaning solutions, such as RNase*Zap*® (Ambion), is recommended. Nuclease-free barrier tips should be used at all times. Always wear gloves and change them often. After tissue homogenization, samples should always be kept on ice. Using nonstick (low retention) RNase-free tubes and tips can be beneficiary when working with low amounts of RNA.

1. Parasite cultures grown in complete RPMI medium at 5% hematocrit (see Note 1).
2. TRIzol® LS Reagent (Invitrogen™) pre-warmed at 37°C.
3. Chloroform.
4. Isopropanol prechilled on ice.
5. Nuclease-free non-DEPC water.
6. DNAse I RNAse-free (Ambion®).
7. Deionized formamide.
8. Formaldehyde, 37%.
9. 10× MOPS EDTA buffer, pH 7.0.
10. Glycerol, 50%.
11. Bromophenol blue powder.
12. Ethidium bromide, 20 mg/mL.
13. GenElute™ mRNA Miniprep Kit (Sigma-Aldrich).
14. 5× RNA storage solution (Ambion).

15. HPLC-purified random hexamers and Anchored OligodT$_{(20)}$.
16. SuperScript® VILO™ cDNA synthesis kit (Invitrogen™).
17. DNA Clean & Concentrator™ (Zymo Research).
18. 5× First Strand Buffer (Invitrogen™): 250 mM Tris–HCl (pH 8.3), 375 mM KCl, 15 mM MgCl$_2$.
19. 5× Second Strand Buffer (Invitrogen™): 100 mM Tris–HCl (pH 6.9), 450 mM KCl, 23 mM MgCl$_2$, 0.75 mM β-NAD$^+$, 50 mM (NH$_4$)$_2$SO$_4$.
20. Set of dATP, dGTP, dCTP, and dUTP.
21. *Escherichia coli* DNA Polymerase I 10 U/μL (Invitrogen™).
22. *E. coli* DNA Ligase 10 U/μL (Invitrogen™).
23. *E. coli* DNA RNase H 2 U/μL (Invitrogen™).
24. 0.1 M DTT.
25. dsDNA Shearase™ (Zymo Research).
26. Encore™ NGS Library System I (NuGEN®).
27. Same-day 70% ethanol in nuclease-free water.
28. USER™ Enzyme (New England Biolabs®).
29. 1× TE buffer, pH 8.0.

3. Methods

3.1. Total RNA Extraction from Parasite Cultures

1. Spin down the cultures at 700 × *g* for 5 min with brake level set at the minimum. Aspirate off the supernatant.
2. Add 5 volumes of pre-warmed TRIzol® LS (37°C) and mix thoroughly to dissolve all clumps (see Note 2).
3. Incubate at 37°C for 5 min to ensure the complete de-proteinization of nucleic acids.
4. *Stopping point*. The samples can be stored at −80°C until further processing. They must be thawed on ice before resuming the protocol.
5. Keep the samples on ice. For each 5 mL of TRIzol® LS that was used in step 2, add 1 mL of chloroform and vortex for 1 min.
6. Centrifuge at 12,000 × *g* for 30 min at 4°C.
7. Carefully transfer the upper aqueous layer to a fresh tube (see Note 3) and add 0.8 volume of prechilled isopropanol to precipitate the RNA. Mix carefully by inverting.
8. *Stopping point*. The tubes can be stored at −20°C overnight until further processing. Their temperature must be equilibrated on ice for a few minutes before resuming the protocol (see Note 4).

9. Mix by inverting and centrifuge at 12,000×*g* for 30 min and 4°C. Carefully aspirate off the supernatant.
10. Allow the pellet to air dry on ice for 5 min and add 30–100 μL of RNase-free non-DEPC-treated water.
11. Heat tubes at 60°C for 10 min and then place on ice.

3.2. DNase Treatment

1. To 100 μL of RNA solution, add 11.3 μL of 10× DNase I Buffer and 2 μL (4 U) of DNase I (2 U/μL).
2. Incubate the tube at 37°C for 30 min.
3. Inactivate the DNAse at room temperature for 5 min in the presence of 1 mM EDTA. Transfer the tube on ice.
4. *Stopping point*. RNA solutions can be aliquoted and stored at −80°C. When needed, thaw tubes on ice. Avoid repeated freeze/thaw cycles.

3.3. Verification of the Quality and the Quantity of Total RNA

1. Quantify the concentration of the total RNA solution by UV spectrophotometry, such as a NanoDrop (Thermo Scientific). Typically, a clean solution of nucleic acid in nuclease-free water has an OD~1.85. A ratio ranging from 1.8 to 2.2 is, therefore, recommended.
2. Check RNA integrity by agarose gel electrophoresis (see Note 5 and Note 6).
3. If genomic DNA is visible on the gel, repeat the DNase treatment (see Note 7).
4. Verify the absence of trace amounts of genomic DNA by 40 cycles of PCR on a chosen control gene using 50–500 ng of total RNA solution. Repeat the DNase treatment if necessary.
5. *Stopping point*. Store the total RNA solution at −80°C.

3.4. Purification of PolyA + mRNA from Total RNA

This protocol uses the reagents from the GenElute™ mRNA Miniprep Kit (Sigma-Aldrich). Before starting, equilibrate a heating block for microcentrifuge tubes at 70°C. Keep the Elution Solution at 70°C. If the beads were kept at 4°C, let them sit on the bench top for at least 15 min. Cold beads reduce yields.

1. Thaw the total RNA sample on ice. The amount of starting material should be 150–500 μg of purified total RNA. The remaining steps are performed at room temperature unless specified otherwise.
2. Adjust volume of total RNA to 250 μL with RNase-free water. Add 250 μL of 2× Binding Solution and vortex briefly.
3. Add 15 μL of Oligo(dT) Polystyrene Beads and vortex thoroughly.

4. Heat the mixture at 70°C for 3 min to denature the RNA and let it cool down for 10 min at room temperature.

5. Centrifuge for 2 min at maximum speed ($14,000–16,000 \times g$) in a tabletop microcentrifuge. Carefully pipette off the supernatant without disturbing the bead pellet (see Note 8).

6. Add 500 μL of Wash Solution and mix by vortexing. Transfer the mixture to a GenElute spin filter/collection tube assembly. Failure to transfer all traces of mixtures will result in lower mRNA yields.

7. Centrifuge for 1 min at maximum speed ($14,000–16,000 \times g$) in a tabletop microcentrifuge. Discard the flow through and place the collection tube back on the GenElute spin filter.

8. Add 500 μL of Wash Solution onto the GenElute spin filter and centrifuge for 2 min at maximum speed. Transfer the GenElute spin filter to a fresh nuclease-free microcentrifuge tube.

9. Add 50 μL of Elution Solution heated at 70°C onto the center of the GenElute spin filter and incubate for 5 min at 70°C. Centrifuge for 1 min at maximum speed.

10. Repeat step 9 for a second elution.

11. Check the mRNA quantity by UV spectrometry. Expect 1.5–2.5% of the starting amount of total RNA, depending on the considered morphological stage of the parasite.

12. *Stopping point*. The mRNA solutions can be stored at −80°C. When needed, thaw tubes on ice. Avoid repeated freeze/thaw cycles.

3.5. Fragmentation of the PolyA + mRNAs

1. Reduce sample volume to 15–20 μL in a vacuum concentrator type SpeedVac® without heating. Do not let the sample dry.

2. Add 4 volumes of 5× RNA storage solution and incubate for 40 min at 98°C (see Note 9).

3. Reduce sample volume to 10 μL in a vacuum concentrator without heating. Do not let the sample dry.

4. *Stopping point*. The mRNA solutions can be stored at −80°C. When needed, thaw tubes on ice. Avoid repeated freeze/thaw cycles.

3.6. First-Strand cDNA Synthesis

First-strand cDNA is synthesized using the SuperScript® VILO™ cDNA synthesis kit (Invitrogen™). All reagents and buffers mentioned in this section refer to elements of the kit. Frozen items should be kept on ice after thawing.

1. In a thin-wall, nuclease-free, 0.2-mL PCR-grade tube, mix 3 μg of random hexamers and 1 μg of Anchored oligodT$_{(20)}$ to the fragmented mRNA in 14 μL final volume (see Note 10).

2. Incubate the tube in a preheated thermal cycler at 70°C for 10 min and quickly chill on ice for 5 min. Do not reduce this time.

3. On ice, add the following reagents to the tube from step 2: 4 µL of 5× VILO™ Reaction Mix and 2 µL of 10× SuperScript® Enzyme Mix (see Note 11). If you prepare multiple samples at the same time, make a master mix containing the 5× VILO™ Reaction Mix and the 10× SuperScript® Enzyme Mix and add 6 µL of it to each sample (see Note 12).

4. Gently mix the sample by flicking the bottom of the tube with fingertips. Spin, and place on ice.

5. Incubate the sample in a thermal cycler using the following program: 25°C for 10 min, 42°C for 90 min, 85°C for 5 min, and hold at 4°C.

6. Remove promptly from the thermal cycler and place the tube on ice.

7. Purify first-strand cDNA using the DNA Clean & Concentrator™ (Zymo Research):

    (a) Add 100 µL of DNA Binding Buffer to the reaction mixture and mix well by pipetting up and down.

    (b) Transfer to a Zymo-Spin™ Column/collection tube assembly and centrifuge for 30 s at maximum speed (14,000–16,000 ×g) in a tabletop microcentrifuge. Discard the flow through.

    (c) Add 200 µL of Wash Buffer (freshly prepared with absolute ethanol, see Note 13). Centrifuge for 30 s at maximum speed.

    (d) Discard the flow through and repeat (c) for a second wash.

    (e) Transfer the Zymo-Spin™ Column to a fresh nuclease-free microcentrifuge tube.

    (f) Pipette 20 µL of nuclease-free water to the column matrix and let stand for 1 min. Centrifuge for 30 s at maximum speed to elute the nucleic acid.

    (g) Repeat step f.

8. Adjust sample volume to 47 µL with non-DEPC nuclease-free water (see Note 14).

3.7. Second-Strand cDNA Synthesis

All reagents and buffers mentioned in this section should be made freshly. Frozen items should be kept on ice after thawing.

1. Prepare a dNTP mix containing dATP, dCTP, dGTP, and dUTP (instead of dTTP), each at 10 mM final concentration (see Note 15).

2. Chill all reagents on ice.

3. Set up the following reaction on ice and in the provided order:

First-strand cDNA	47 μL
5× First Strand Buffer	2 μL
100 mM DTT	1 μL
5× Second Strand Buffer	15 μL
10 mM dNTP (w/dUTP) mix	4 μL
E. coli DNA Polymerase I, 10 U/μL	4 μL
E. coli DNA Ligase, 10 U/μL	1 μL
E. coli RNase H, 2 U/μL	1 μL

4. Mix gently by pipetting and incubate at 16°C for 2 h.
5. Chill the reaction on ice for at least 5 min.
6. Purify double-stranded (ds) cDNA using the DNA Clean & Concentrator™ (Zymo Research):
   (a) Add 375 μL of DNA Binding Buffer to the reaction mixture and mix well by pipetting up and down.
   (b) Transfer to a Zymo-Spin™ Column/collection tube assembly and centrifuge for 30 s at maximum speed (14,000–16,000 × *g*) in a tabletop microcentrifuge. Discard the flow through.
   (c) Add 200 μL of Wash Buffer (freshly prepared with absolute ethanol, see Note 13). Centrifuge for 30 s at maximum speed.
   (d) Discard the flow through and repeat step c for a second wash.
   (e) Transfer the Zymo-Spin™ Column to a fresh nuclease-free microcentrifuge tube.
   (f) Pipette 6 μL of nuclease-free water to the column matrix and let stand for 1 min. Centrifuge for 30 s at maximum speed to elute the nucleic acid.
   (g) Repeat step f.
7. Check the ds cDNA quantity by UV spectrometry and quality by visualization on a 1.2% agarose gel electrophoresis. A smear should be easily detected (see Note 16).
8. *Stopping point*. The sample is now ds cDNA and is relatively stable. It can be stored at −20°C. When needed, thaw tubes on ice. Avoid repeated freeze/thaw cycles.

3.8. ds cDNA Fragmentation

1. Mix 700 ng of ds cDNA with 11.5 μL of 3× dsDNA Shearase™ Reaction buffer and 3.5 μL of dsDNA Shearase™ (Zymo Research). Reach a final volume of 35 μL with nuclease-free water.

2. Incubate at 37°C for 40 min (see Note 17).

3. Purify ds cDNA and inactivate the dsDNA Shearase™ by adding 175 μL of the DNA Clean & Concentrator™ DNA binding buffer (Zymo Research).

4. Mix well by pipetting up and down, transfer to a Zymo-Spin™ column/collection tube assembly, and centrifuge for 30 s at maximum speed ($14{,}000\text{--}16{,}000 \times g$) in a tabletop microcentrifuge. Discard the flow through.

5. Add 200 μL of Wash Buffer (freshly prepared with absolute ethanol, see Note 13). Centrifuge for 30 s at maximum speed.

6. Discard the flow through and repeat step 5 for a second wash.

7. Transfer the Zymo-Spin™ column to a fresh nuclease-free microcentrifuge tube.

8. Pipet 10 μL of nuclease-free water to the column matrix and let stand for 1 min. Centrifuge for 30 s at maximum speed to elute the nucleic acid.

9. Repeat step 8.

10. Check the size range and the concentration of the sample using microfluidic-based separation devices suitable for small amounts of starting materials, such as an Agilent 2100 Bioanalyzer (Agilent Technologies) or a LabChip® GX (Caliper Life Sciences) (see Note 18). Repeat the fragmentation procedure if necessary.

11. *Stopping point.* The sample can be stored at −20°C. When needed, thaw tubes on ice. Avoid repeated freeze/thaw cycles.

3.9. Library Preparation

The protocol described here uses the NuGEN® Encore™ NGS Library System I, compatible with the Illumina® Genome Analyzer and Hi-Seq sequencing platforms, and all mentioned reagents refer to components of this kit (see Note 19). However, since the starting material is double-stranded DNA, it can be easily adapted to any gDNA library preparation kit (including multiplexing) or set of reagents.

The Agencourt® magnetic beads used for sample cleanup must be incubated at room temperature for at least 15 min before use. Cold beads reduce yields. Before each use, beads must be fully resuspended by inverting and tapping the tube. Thaw all necessary reagents, mix by vortexing, spin, and keep them on ice until use. Keep the nuclease-free water at room temperature.

3.9.1. End Repair

1. Dilute 200 ng of fragmented ds cDNA to a volume of 7 μL with nuclease-free water in a 0.2-mL, thin-wall, nuclease-free PCR tube. Place on ice.

2. On ice, add 2.5 μL of End Repair Buffer Mix and 0.5 μL of End Repair Enzyme Mix to the sample and mix by pipetting

up and down. If more than one sample is treated, prepare a master mix of sufficient amounts of End Repair Buffer Mix and End Repair Enzyme Mix before adding 3 μL to each sample (see Note 12).

3. Place the tube in a pre-warmed thermal cycler (lid heated at 100°C) with the following program: 30 min at 25°C; 10 min at 70°C; hold at 4°C.
4. Remove the sample promptly from the thermal cycler, give a quick spin, and place on ice.
5. Resuspend Agencourt® RNAClean XP magnetic beads by inverting and tapping the tube on the bench top. Do not spin the tube.
6. Add 12 μL of the bead slurry to the sample and mix thoroughly by pipetting up and down. Incubate at room temperature for 10 min.
7. Transfer tubes to the magnetic separation device and let them stand for 5 min (see Note 20).
8. While still on the magnet, carefully pipette off 15 μL of liquid without disturbing the beads (see Note 21). Dispersion and loss of significant amounts of beads will reduce yields.
9. While still on the magnet, gently add 200 μL of freshly made 70% ethanol and let stand for 30 s (see Note 22).
10. While still on the magnet, remove 200 μL of the ethanol wash (see Note 23).
11. Repeat step 9.
12. While still on the magnet, remove all of the ethanol wash. Carefully inspect the tube for the absence of ethanol drops.
13. While still on the magnet, air dry the beads for 5–10 min. Carefully inspect the tube to ensure that the ethanol has entirely evaporated.
14. Remove from the magnet and add 12 μL of nuclease-free water to the dried beads. Resuspend carefully by pipetting up and down.
15. Transfer the tubes to the magnet and let them stand for 1 min.
16. While on the magnet, carefully remove 11 μL of the eluate without disturbing the beads and transfer to a fresh tube.
17. Repeat step 15 to minimize the carryover of beads into the next stage of the library preparation.
18. While on the magnet, carefully remove 10 μL of the eluate without disturbing the beads and transfer to a fresh, nuclease-free, thin-wall, 0.2-mL PCR tube. Place on ice.
19. Proceed immediately to Subheading 3.9.2 (see Note 24).

3.9.2. Ligation

1. On ice, add 1 μL of Adaptor Mix to the sample (see Note 25). Mix by pipetting thoroughly with the pipette set to 5 μL.

2. On ice, add 12.5 μL of Ligation Buffer Mix and 1.5 μL of Ligation Enzyme Mix to the sample (the Ligation Buffer Mix is very viscous and should be pipetted slowly). If more than one sample is treated, prepare a master mix of sufficient amounts of Ligation Buffer Mix and Ligation Enzyme Mix before adding 14 μL to each sample (see Note 12).

3. Carefully mix by pipetting slowly up and down without forming bubbles with the pipette set at 20 μL. Spin down the tube for 2 s.

4. Place the tube in a pre-warmed thermal cycler (lid not heated) with the following program: 10 min at 25°C; hold at 4°C. IMPORTANT: Use this incubation time to prepare the Amplification Master Mix to be used in the library amplification reaction (see Subheading 3.9.3, step 1). The adapter-ligated sample must not remain on ice for more than 10 min from the end of the ligation reaction to the beginning of the amplification reaction.

5. Remove the sample promptly from the thermal cycler, give a quick spin, and place on ice.

6. Proceed immediately to Subheading 3.9.3.

3.9.3. Library Amplification

1. Prepare an Amplification Master Mix by sequentially mixing the following reagents: 64 μL of Amplification Buffer Mix, 3 μL of Amplification Primer Mix, and 4 μL of DMSO (this mix should have been prepared during the incubation indicated at Subheading 3.9.2, step 4). Place tube on ice. If more than one sample is treated, adapt volumes to prepare a sufficient quantity of master mix.

2. On ice, add 3 μL of Amplification enzyme mix and 1 μL of USER™ enzyme to the Amplification Master Mix immediately before adding to the adapter-ligated sample (the USER™ enzyme will degrade the second strand of the ds cDNA prior to amplification to achieve strand specificity, see Note 26). If more than one sample is treated, adapt volumes to prepare a sufficient quantity of master mix.

3. Mix well by pipetting slowly, avoiding bubbles, spin, and place on ice.

4. Add 73 μL of Amplification Master Mix to a clean, 0.2-mL, thin-wall, nuclease-free PCR tube.

5. Add 7 μL of adapter-ligated sample to the tube prepared at step 4. Mix well by pipetting slowly up and down at the 73 μL pipette setting, avoiding bubbles, spin, and place on ice. The remaining adapter-ligated sample can be discarded.

6. Place the tube in a pre-warmed thermal cycler (lid heated at 100°C) with the following program: 5 min at 95°C; 2 min at 72°C; 5 cycles of 30 s at 94°C—30 s at 55°C—1 min at 72°C; 10 cycles of 30 s at 94°C—30 s at 63°C—1 min at 72°C; 5 min at 72°C; hold at 10°C.

7. Remove the sample promptly from the thermal cycler, give a quick spin, and place on ice.

8. Resuspend Agencourt® RNAClean XP magnetic beads by inverting and tapping the tube on the bench top. Do not spin the tube.

9. Add 80 μL of the bead slurry to the amplified library and mix thoroughly by pipetting up and down (see Note 27). Incubate at room temperature for 10 min.

10. Transfer tubes to the magnetic separation device and let them stand for 5 min (see Note 20).

11. While still on the magnet, carefully pipette off 140 μL of liquid without disturbing the beads (see Note 21). Dispersion and loss of significant amounts of beads will reduce yields.

12. While still on the magnet, gently add 200 μL of freshly made 70% ethanol and let stand for 30 s (see Note 22).

13. While still on the magnet, remove 200 μL of the ethanol wash (see Note 23).

14. Repeat steps 12–13 two more times for a total of three washes.

15. While still on the magnet, remove all of the ethanol wash. Carefully inspect the tube for the absence of ethanol drops.

16. While still on the magnet, air dry the beads for 10–15 min. Carefully inspect the tube to ensure that the ethanol has entirely evaporated.

17. Remove from the magnet and add 33 μL of 1× TE to the dried beads. Resuspend carefully by pipetting up and down.

18. Transfer the tubes to the magnet and let stand for 2 min.

19. While on the magnet, carefully remove 30 μL of the eluate without disturbing the beads and transfer to a fresh tube. Place on ice.

20. *Stopping point*. The amplified libraries can be stored at −20°C. When needed, thaw tubes on ice. Avoid repeated freeze/thaw cycles.

3.9.4. Qualitative and Quantitative Evaluation of the Library

1. Analyze 3 μL of the library on a 1.6% agarose gel electrophoresis (see Note 28) and check the size and the purity of the library. Quantify by UV spectrophotometry.

4. Notes

1. Typically, parasites are cultured in 25 mL total volume at 5% hematocrit until a parasitemia of 6–10% is reached. If a synchronization is performed (e.g., using sorbitol), make sure to let the parasites recover from the stress of the treatment, and ideally wait for one cycle of invasion, before harvesting the RNAs. Waiting will minimize the background caused by stress-related variations. This RNA-seq protocol is typically prepared using four different cultures pooled together.

2. It is crucial to dissolve everything at this step for an optimal yield.

3. Do not transfer any of the lower phase to the next step. Phenol inhibits downstream enzymatic reactions, including reverse transcription.

4. Tubes containing nucleic acids that were precipitated at −20 or −80°C should always be allowed to equilibrate on ice before centrifugation. At these low temperatures, the samples tend to become very viscous and the efficiency of centrifugation is lower.

5. Typically, 0.5–1 µg of total RNA should be loaded on a 1.2% agarose gel. Mix sample with 10 volumes of denaturing RNA loading buffer (for 1.5 mL stock loading buffer mix: 750 µL of deionized formamide, 240 µL of formaldehyde, 37%, 150 µL of 10× MOPS EDTA buffer, pH 7.0, 200 µL of 50% glycerol, 0.5 mg of bromophenol blue, and 10 µL of ethidium bromide, 10 mg/mL) and heat for 5 min at 65°C. Ensure that all solutions and hardware, including electrophoresis tank and gel combs, are RNAse free. The 28- and 18-S rRNAs should appear as two clean bands around 5.3 and 2 kb, respectively. The upper band should be more intense. The presence of significant smearing or a lower intensity of the upper band indicates degradation of the extracted material.

6. If the purity of the RNA solution is questioned, e.g., presence of phenol, the samples can be further purified on RNeasy® (QIAGEN) clean-up columns according to the "RNA Cleanup" manufacturer's protocol. All solutions must be fresh.

7. It is crucial to eliminate all contamination with genomic DNA in order to avoid competition in downstream reaction and inaccurate quantitative analysis of RNA levels or false discovery of alternative transcripts.

8. Any loss in beads will result in a loss of material. For maximum yield, about 50 µL of sample should remain in the tube after removing the supernatant.

9. The efficiency of this step is directly linked to the amount of starting material. If desired, the incubation time can be adjusted accordingly but should not exceed 60 min.

10. A combination of random primers and oligos dT should always be used in experiments dealing with *P. falciparum*'s AT-rich genome to maximize the reverse transcription of all possible transcripts regardless of their GC content.

11. The 5× VILO™ Reaction mix already contains random primers, $MgCl_2$, and dNTPs. The 10× SuperScript® Enzyme Mix includes the SuperScript® III Reverse Transcriptase (reduced RNase H activity and high thermal stability for extended synthesis), the RNAseOUT™ Recombinant Ribonuclease Inhibitor, and a helper protein proprietary to Invitrogen™.

12. When dealing with multiple samples at the same time, the delay between the preparation of the first sample and the preparation of the last sample should be kept to a minimum to ensure uniformity. Do not prepare more than eight samples at a time.

13. As a general rule, when using nucleic acid cleanup and purification reagents, buffers containing ethanol should always be as fresh as possible. Aging solutions can cause dramatic losses in material.

14. At that point, the samples can theoretically be frozen at −20°C until further processing. Empirical observations seem to indicate, however, that the performances are significantly increased when second-strand cDNA is synthesized immediately after first strand. Therefore, we do not recommend the freezing of first-strand cDNA.

15. The substitution of the dTTP by dUTP in the dNTP mix is critical in this protocol since it allows using the USER™ (Uracil-Specific Excision Reagent) enzyme prior to library amplification and achieving strand specificity. The USER™ enzyme will leave a nucleotide gap at the location of a uracil in the second strand of the cDNA.

16. Obtaining high-quality ds cDNA is an absolute prerequisite for a successful preparation of a sequencing library. We recommend not proceeding if the ds cDNA is not of satisfactory quality (the presence of a regular smear on the gel and an $OD \geq 1.8$ is an example of satisfactory quality).

17. These reaction conditions have been optimized to obtain fragments ranging 150–300 bp in size. Increase incubation time for shorter fragments, and decrease it for longer ones.

18. Small amounts of nucleic acids cannot be detected by classical agarose gel electrophoresis. In order to avoid wasting large amounts of samples, we recommend using microfluidic-based

devices that can quantify and display the size distribution of a few microliters of a sample concentrated in the picogram per microliter range. The Bioanalyzer DNA High Sensitivity Chip (Agilent Technologies) can resolve 3 μL of purified DNA at 5 μg/μL in TE for sizes ranging 50–7,000 bp. The LabChip® GX can resolve bands as low as 5 bp and features a sensitivity of 0.1 ng/μL.

19. The NuGEN® Encore™ NGS Library System I uses magnetic beads (RNAClean® XP Purification Beads supplied in the kit) for the successive purification of the samples through the library preparation steps rather than silicate-based spin columns. Magnetic beads allow for minimized sample loss and reduction of input material for library preparation. A magnetic separation device, such as the Agencourt® SPRIStand, is thus necessary to perform the purification steps. When using the Agencourt® SPRIStand, 96-well plates or tube strips are preferred rather than single tubes for greater stability in the stand and better separation.

20. Reduction in the incubation time of the beads on the magnetic stands will result in reduced recovery of the samples. Similarly, the various incubation times have been optimized to obtain reproducible results in terms of nucleic acid yield and size range. They must be strictly observed.

21. While on the magnet, the beads will stay on the walls of the tube and form a ring. Use a small-volume pipette tip to reach the bottom of the tube without touching the sides and gently aspirate the desired volume.

22. If multiple samples are treated simultaneously, monitor the time spent in reaching the last tube and deduct it from the 30 s.

23. Always use the smallest volume pipette tip that allows the removal of the desired volume within 2–3 withdrawals. Do not try to get everything in one step and pipette slowly to prevent any bead loss.

24. NuGEN® developed proprietary adapter and primer sequences directly compatible with the Illumina® Genome Analyzer and Hi-Seq systems. Its use differs from the more common Illumina® ones mostly in the fact that the step for 3′-end A-tailing of the fragments that is usually carried on prior to adapter ligation is absent in the NuGEN® protocol. This specificity significantly reduces the hands-on time of the protocol. In addition, NuGEN® adapters generate libraries free from adaptor dimers, unlike Illumina®'s adapters.

25. The adapters are partly complementary and provided partially annealed to each other. This condition is necessary for a successful ligation to the sample of interest. Make sure to always keep the tube of Adapter Mix on ice so that the adapter duplex does not denature.

26. The USER™ enzyme is added to the amplification mix. During a short denaturing step (5 min at 95°C) prior to the actual amplification cycles, the USER™ enzyme nicks the second strand of the ds cDNA at uracil locations. Only the first strand is amplified and strand specificity is achieved.

27. If multiple samples are processed at the same time, it may be useful to use a multichannel pipette to ensure consistent incubation times.

28. A high percentage of agarose is necessary to resolve small libraries. Increasing the amount of agarose, however, significantly increases the detection threshold using intercalant agents, such as ethidium bromide. Here, preparing a gel at 1.5–1.8% agarose is a good compromise between resolution and sensitivity. In addition, the use of low-range agarose, such as the Certified Low Range Ultra Agarose (Bio-Rad), greatly improves the resolution of small bands without having to increase the agarose content.

Acknowledgments

The authors thank Courtney Brady (NuGEN®), and Barbara Walter, John Weger, Rebecca Sun, and Glenn Hicks (Institute for Integrative Genome Biology, University of California Riverside) for their assistance in the library preparation and sequencing processes.

References

1. Le Roch KG et al (2003) Discovery of gene function by expression profiling of the malaria parasite life cycle. Science 301(5639):1503–1508
2. Bozdech Z et al (2003) The transcriptome of the intraerythrocytic developmental cycle of *Plasmodium falciparum*. PLoS Biol 1(1):E5
3. Otto TD et al (2010) New insights into the blood-stage transcriptome of *Plasmodium falciparum* using RNA-Seq. Mol Microbiol 76(1):12–24
4. Katherine Sorber MT, Dimon JLDeRisi (2011) RNA-Seq analysis of splicing in *Plasmodium falciparum* uncovers new splice junctions, alternative splicing and splicing of antisense transcripts. Nucleic Acids Res 39(9): 3820–3835

Chapter 5

RNA In Situ Hybridization in Arabidopsis

Miin-Feng Wu and Doris Wagner

Abstract

RNA in situ hybridization using digoxigenin-labeled riboprobes on tissue sections is a powerful technique for revealing microscopic spatial gene expression. Here, we describe an in situ hybridization method commonly practiced in *Arabidopsis* research labs. The highly stringent hybridization condition eliminates the usage of Ribonlucease A and gives highly specific signals. This also allows the use of longer probes which enhance signal strength without cross hybridization to closely related genes. In addition, using spin columns in template and riboprobe purification greatly reduces background signals.

Key words: In situ hybridization, FAA, Paraplast, Sections, Nonradioactive, DIG, Alkaline phosphatase, NBT/BCIP

1. Introduction

Studying spatial mRNA transcript accumulation patterns provides valuable information of how gene function is correlated with morphological events during development. In *Arabidopsis*, reporter assays and RNA in situ hybridization are commonly used in studying the spatial gene expression patterns in tissues. However, despite the ease of making transgenic *Arabidopsis* carrying the reporter gene constructs, gene silencing is common among transgenic lines. In addition, reporter gene expression patterns are affected by the position where the transgene is inserted in the genome. Scientists oftentimes have to screen through multiple transgenic lines to find a representative expression pattern. Moreover, the reporter constructs often miss critical *cis*-regulatory elements or do not reveal posttranscriptional regulatory events, including miRNA-mediated transcript cleavage. RNA in situ hybridization enables scientists to directly study cell-specific gene expression in tissue sections without generating multiple transgenic lines. This method allows scientists to compare a given gene's expression profile among various genotypes in one experiment. It can further distinguish the tissue-specific differential gene expression levels among genotypes by comparing their tissue sections on the same slide. Since RNA in situ hybridization reveals mRNA transcript accumulation combining all inputs, it can serve as a gold standard for spatial mRNA expression patterns of the gene of interest.

Here, we describe an RNA in situ hybridization method using digoxigenin (DIG)-labeled riboprobes on paraplast sections. Using the stable, nonradioactively labeled probes and colorimetric detection of hybridization sites solves previous issues with the poor image resolution in radioactive in situ hybridization. Scientists can easily distinguish cell-specific expression patterns with differential interference contrast (DIC) microscopy. In this protocol, we used highly stringent hybridization conditions and eliminated the Ribonuclease A (RNase A) digestion after hybridization. The combination of a longer probe length and a higher hybridization temperature in 50% formamide-containing hybridization buffer allows us to distinguish similar genes without cross hybridization (1). This method has been proven to be effective for most of *Arabidopsis* tissues. With slight modifications, it should be applicable to other plant species. This protocol is developed based on modifications of previously published methods and by different laboratories (2, 3).

2. Materials

Diethylpyrocarbonate (DEPC)-treated water is only necessary for fixation, probe synthesis, and in the steps noted. Otherwise, autoclaved ultrapure water is sufficient. Dispose formaldehyde, formamide, and histoclear properly according to waste disposal regulations. Do all the steps involving formaldehyde in a chemical fume hood.

2.1. Fixation/Dehydration/Embedding/Sectioning

1. Fixative (FAA): 50% ethanol, 3.7% formaldehyde, 5% acetic acid. Prepare freshly using DEPC-H_2O right before fixation.
2. 20-ml Glass scintillation vials.
3. Eosin Y.
4. Histoclear.
5. Paraplast plus tissue embedding medium.
6. Embedding hot plate.
7. Aluminum weighing dish.
8. Disposable base molds.

9. Microtome and slide warmer.
10. ProbeON plus slides (Fisher Scientific).

2.2. In Vitro Transcription

1. 10× DIG RNA labeling mix (Roche).
2. SP6 and T7 RNA polymerases, transcription buffers, 100 mM DTT, and RNase inhibitor.
3. RQ1 RNase-free DNase.
4. DNA purification spin columns and RNA purification spin columns.
5. tRNA (20 mg/ml).
6. 2× Carbonate (CO_3) hydrolysis buffer: 300 µl 200 mM Na_2CO_3 and 200 µl 200 mM $NaHCO_3$. Prepare 200 mM Na_2CO_3 and 200 mM $NaHCO_3$ freshly with DEPC-H_2O before use.
7. 10% Acetic acid: Prepared with DEPC-H_2O.
8. 3 M Sodium acetate, pH 5.2. Treat with DEPC overnight and autoclave.

2.3. In Situ Section Pretreatment/Hybridization

1. 20× SSC: 3 M NaCl, 300 mM sodium citrate. Adjust pH to 7.0 and autoclave.
2. 1 M Tris–HCl, pH 8.0. Adjust pH to 8.0 and autoclave.
3. 500 mM EDTA, pH 8.0. Adjust pH to 8.0 and autoclave.
4. 10× PBS: 1.3 M NaCl, 70 mM Na_2HPO_4, 30 mM NaH_2PO_4. Adjust pH to 7.0 and autoclave.
5. PCR grade recombinant proteinase K (Roche).
6. Glycine.
7. 2 M Triethanolamine, pH 8.0. Do not autoclave triethanolamine.
8. Acetic anhydride.
9. 10× In situ salt: 3 M NaCl, 100 mM Tris–HCl, pH 8.0, 50 mM EDTA, 100 mM sodium phosphate, pH 6.8. Store at room temperature.
10. 50× Denhardt's solution (Sigma-Aldrich).
11. Ultrapure deionized formamide for hybridization buffer. Formamide ≥99% (GC) for soaking paper towels.
12. Hybridization buffer: 0.11 g dextran sulfate, 253 µl DEPC-H_2O, 110 µl 10× in situ salts, 440 µl ultrapure deionized formamide, 22 µl 50× Denhardt's solution, 55 µl tRNA (20 mg/ml). The total volume is 880 µl, enough for five slides. Add dextran sulfate to DEPC-H_2O in a microcentrifuge tube. Leave the tube on 75°C heat block to dissolve dextran sulfate. Vortex periodically during section pretreatment. The solution will look cloudy when the dextran sulfate is dissolved. Subsequently,

let the solution cool to room temperature and add the 10× in situ salts, ultrapure deionized formamide, 50× Denhardt's solution, and tRNA. Mix well. Transfer to a 65°C heat block when ready to do hybridization.

13. Cover glass, 124 × 60 mm.

2.4. Post-hybridization

1. 10× TBS: 1 M Tris–HCl, pH 7.5, 1.5 M NaCl.
2. Square plastic boxes, 4.3 × 4.3 × 1.125 in. (L × W × H).
3. Blocking solution: 1% Blocking Reagent (Roche) in 1× TBS. Prepare blocking solution while waiting for the 0.2× SSC washes. Microwave 1× TBS to 65–75°C. Add Blocking Reagent powder to heated 1× TBS. Stir vigorously for 5–10 min to dissolve. Let the solution cool on ice before adding it to slides.
4. TBX: 1× TBS, 1% BSA (globulin-free, protease-free, Sigma-Aldrich A3059), 0.3% Triton X-100. Prepare freshly before use. Store TBX at 4°C during antibody incubation.
5. Anti-DIG-AP, Fab fragments (Roche).
6. Detection Buffer: 100 mM Tris–HCl, pH 9.5, 100 mM NaCl, 50 mM $MgCl_2$. Prepare freshly before use.
7. Nitro blue tetrazolium chloride/5-Bromo-4-chloro-3-indolyl phosphate, toluidine salt (NBT/BCIP) stock solution (Roche) or Western Blue (Promega).
8. 1 M Levamisole ((−)-Tetramisole hydrochloride). Dissolved in water. Store at −20°C.
9. Mounting medium. Hydrophobic: Permount (Fisher Scientific). Aqueous: Aqua-Mount (Lerner Laboratories).

3. Methods

3.1. Fixation

1. Aliquot 10 ml of FAA into a 20-ml glass scintillation vial.
2. Harvest tissues into FAA. Keep the vial on ice while harvesting the tissue.
3. The total fixation time is 2 h 30 min. At the start of fixation, vacuum infiltrate for 15 min at room temperature to pull out the air trapped in the tissues and allow better penetration of the fixative. Repeat the vacuum step until the tissues sink or for at least 45 min (see Note 1). The time during vacuum infiltration is counted toward the total fixation time.
4. Put the vial on a rotary platform at room temperature until 2 h 30 min is reached.

3.2. Dehydration

1. Prepare the ethanol series freshly before each step. Make 10–15 ml per sample.

   After fixation, remove FAA by aspiration and replace with 50% ethanol. Incubate at room temperature for 30 min on a rotary platform. Continue dehydrating the tissues through the following ethanol series: 50, 60, 70, 80, and 90% for 30 min each at room temperature on a rotary platform. The tissues can be stored in 70% ethanol for several months at 4°C. After the 90% ethanol step, replace with 95% ethanol plus 0.1% of Eosin Y and leave the tissues overnight at 4°C.

2. Continue dehydrating the tissues with three changes of 100% ethanol for 30 min each on a rotary platform at room temperature.

3. Successively replace the ethanol with the following solutions for 30–60 min each step:

   25% Histoclear/75% Ethanol

   50% Histoclear/50% Ethanol

   75% Histoclear/25% Ethanol

   3× 100% Histoclear

4. After the last histoclear step, remove histoclear and replace it with 10 ml fresh histoclear. Add 15–20 paraplast chips and move the vial to a 60°C incubator. Continue adding 15–20 paraplast chips every 30 min until the vial is full. Leave the tissues in this 50% histoclear/50% paraplast mixture overnight at 60°C.

5. Do six wax changes over the next 2 days (three times per day, 4 h apart between each wax change) (see Note 2).

3.3. Embedding

1. Turn on the embedding hot plate 1–2 h before starting.

2. Put an aluminum dish at the very end on the hottest part. Take out the vial from 60°C. Swirl the vial and dump the tissues quickly into the aluminum dish.

3. Pour molten wax into a plastic base mold. Place the tissue into the base mold and arrange it either parallel (for longitudinal sections) or perpendicular (for transverse sections) to the bottom of base mold.

4. Move the base mold to the coolest part of the hot plate right away.

5. After the wax is hardened completely, stack the base molds and store them in a Tupperware box with some Drierite on the bottom at 4°C (see Note 3).

3.4. Sectioning

1. Before beginning, remove embedded tissues from 4°C and let them warm to room temperature (see Note 4).

2. Remove the sample from the base mold. Trim the wax block roughly and mount it onto a microtome block. Make sure that the tissue is either parallel (for longitudinal sections) or perpendicular (for transverse sections) to the base of the microtome block.

3. Under a dissection microscope, further trim the wax block into a trapezoid. The top of the trapezoid must be parallel to the bottom.

4. Set the slide warmer to 42°C. Place a ProbeOn Plus slide onto the slide warmer. Add 2 ml of DEPC-H_2O to the slide.

5. Section tissues at 8 μm.

6. Cut the resulting wax ribbons into 2-cm sections with a razor blade. Transfer sections to the top of water on the slide by forceps. Adjust the position with an artist paintbrush. The first section should go all the way to the side away from the slide-labeling region. Place the next section right up against the previous one. Once all the sections are added, remove water using a P1000 Pipetman. Absorb excess water with a damp Kimwipe. Leave the sections on the slide warmer overnight to dry. The dried slides can be stored in slide boxes with Drierite at 4°C for 2–3 days.

3.5. In Vitro Transcription of DIG-Labeled Riboprobes

1. Clone your probe in plasmids with T7 or SP6 RNA polymerase promoters on either side of the multiple cloning sites (see Note 5).

2. Linearize 5 μg of the template plasmid DNA using restriction enzymes with 5′ overhang. Avoid enzymes that generate 3′ overhangs (see Note 6).

3. Clean up the linearized template using DNA purification spin columns. After DNA is bound to the column, wash the column two more times with the binding buffer to eliminate any protein contamination. Wash the column three times with the wash buffer to get rid of residual salt contamination. Elute DNA from the column according to manufacturer's instructions (see Note 7).

4. Set up in vitro transcription reactions. Add the following reagents at room temperature (see Note 8) in the order listed:

   Total volume 25 μl

   10.5 μl 2 μg Linearized template DNA+DEPC-H_2O

   5 μl 5× Transcription buffer

   2.5 μl 100 mM DTT

   2.5 μl 10× DIG RNA labeling mix

   1.5 μl RNase Inhibitor

3 μl T7 or SP6 RNA Polymerase

Incubate at 37°C for 2 h

5. Add an additional 1 μl of RNA polymerase to boost probe yields. Incubate at 37°C for another 30 min.

6. Remove 1 μl from the transcription reaction and save it on ice for gel electrophoresis.

7. Add 73 μl of DEPC-H$_2$O and 2 μl of RNase-free DNase to the remainder. Incubate at 37°C for 10 min.

8. Clean up the transcription reaction using RNA purification spin columns (see Note 9). Elute the probes to 100 μl of RNase-free water. Save 5 μl from the probes and leave it on ice. Add 5 μl of tRNA (20 mg/ml) to the remaining 95 μl of probes.

9. Run a 1.2% 0.5× TBE gel to check the yield of probes from steps 6 and 8 at 90 V for 15 min (see Note 10).

10. Hydrolyze the probes to 150 bp to allow better tissue penetration (4) by incubating in carbonate (CO_3) buffer at 60°C. The length of incubation is determined by the following formula:

    Time (minutes) = (Li-Lf)/(K)(Li)(Lf).

    Li = Initial length of probe (in kbp).

    Lf = Final length of probe (0.15 kbp).

    K = 0.11.

11. Add 100 μl of 2× CO_3 buffer to the probes. Incubate at 60°C for the calculated time.

12. Neutralize the reaction by adding the following in the order listed:

    10 μl 10% Acetic acid

    21 μl 3 M Sodium acetate (pH 5.2)

    460 μl 100% Ethanol

13. Precipitate the probes at −80°C for at least 15 min. Spin down the pellet and wash it with 70% ethanol. Resuspend the pellet in 50 μl of DEPC-H$_2$O and store it at −80°C.

3.6. In Situ Section Pretreatment

3.6.1. Preparation

1. Before the hybridization day, clean and bake the glass staining dishes at 200°C for at least 8 h. Autoclave ultrapure water.

2. On the hybridization day, before starting the pretreatment, prepare the ethanol series and 1× PBS. Dissolve dextran sulfate for the hybridization solution. Prewarm the proteinase K buffer at 37°C.

3. All steps are at room temperature unless otherwise noted.

3.6.2. Section Pretreatment

1. Remove wax and rehydrate sections by the following steps:

   2× 15 min histoclear (see Note 11)

   2× 2 min 100% ethanol

   2 min 95% ethanol

   2 min 70% ethanol

   2 min 50% ethanol

   2 min 30% ethanol

   2× 2 min H_2O

2. Continue with the following steps:

   20 min 2× SSC (see Note 12)

   30 min proteinase K (1 µg/ml) in 100 mM Tris–HCl, pH 8.0, and 50 mM EDTA (see Note 13)

   2 min 2 mg/ml glycine in 1× PBS

   2× 2 min 1× PBS

   10 min 3.7% formaldehyde in 1× PBS (see Note 14)

   2× 5 min 1× PBS

   10 min 0.1 M triethanolamine (pH 8.0) and 0.5% acetic anhydride (see Note 15)

   2× 5 min 1× PBS

3. Dehydrate the sections through the ethanol series from step 1 in reverse order: starting from 30% and ending in 100% ethanol.

4. Dry the slides for 1–2 h under vacuum (see Note 16).

3.7. Hybridization

3.7.1. Preparation

Set up several humidified boxes by placing several layers of paper towels that are soaked with 50% formamide and 0.3 M NaCl at the bottom of the boxes. Put two plastic rods or two 10-ml plastic pipets in the box for elevating the slides. Prewarm the boxes to 53°C.

3.7.2. Hybridization

1. After dextran sulfate is dissolved, add the remaining components of the hybridization buffer. Keep the hybridization buffer at 65°C (see Note 17).

2. Aliquot different amounts of probes into separate microcentrifuge tubes (see Note 18). Bring the volume to 20 µl with DEPC-H_2O, and add 20 µl of ultrapure deionized formamide to the probes.

3. Denature the probes at 80°C for 2 min. Put them on ice right after denaturing.

4. Add 160 µl of prewarmed hybridization buffer to the probe. Mix gently and avoid air bubbles. The final volume for the hybridization solution is 200 µl. Leave the hybridization solution at 65°C.

5. Apply the hybridization solution to one end of the slide. Slowly place down a coverslip. Avoid air bubbles and make sure that the hybridization solution is evenly distributed over the sections. Place slides in the humid boxes and hybridize overnight at 53°C.

3.8. Post-hybridization Washes

1. Float off the coverslips in a small plastic tray with 65°C 2× SSC. Place slides in square plastic boxes with 0.2× SSC (see Note 19). Wash slides four times in 0.2× SSC for 30 min, each at 53°C with gentle agitation.

2. Incubate slides in 1× PBS for 5 min or leave slides in 1× PBS overnight at 4°C.

3.9. Blocking and Anti-DIG Antibody Binding

1. All the steps are at room temperature. Except for the antibody binding step, gently agitate the slides in square plastic boxes on a rotary platform for all steps.

2. Block and equilibrate the slides for antibody binding by the following steps:

   45 min Blocking solution

   45 min TBX

3. Dilute anti-DIG antibody 1:1,250 in TBX. Prepare antibody solution 30 min before the antibody binding step and leave it on ice.

4. Prepare humid boxes with paper towels soaked with water at the bottom. Put two plastic rods or two 10-ml plastic pipets to elevate the slides.

5. Take out slides from TBX. Wipe off excess liquid on the back of the slide. On an elevated platform, apply 100–200 µl of antibody to the slide. Gently place down a coverslip. Avoid air bubbles. Place the slides in the humid boxes for 2 h.

6. Float off coverslips in TBX as in step 1. Wash the slides four times in TBX for 15 min each.

3.10. Detection

1. Wash the slides two times with detection buffer for 10 min each.

2. Prepare alkaline phosphatase substrate solution from the concentrated NBT/BCIP stock according to manufacturer's instruction (see Note 20).

3. Wipe off excess liquid on the back of the slide. Apply 100–200 µl of substrate solution to the slide. Carefully put down a coverslip and avoid air bubbles. Transfer the slide to a humid box. Seal the boxes with parafilm and leave them in darkness for 1–3 days (see Note 21).

4. Float off the cover glass in TE buffer in a small plastic tray to terminate the color reaction. Slides can be stored in TE

buffer for a day or two at this point. Dehydrate slides through ethanol series (see Note 22) if hydrophobic mounting medium is used, or directly mount the slides in aqueous mounting medium.

4. Notes

1. Most types of *Arabidopsis* tissue sink after two vacuum infiltrations.
2. The purpose of this step is to replace the histoclear with wax. Do not cover the vials with lids; otherwise, histoclear cannot evaporate. When pouring wax, let the wax at the bottom set slightly by placing the vials on your hand so that the tissues do not pour out.
3. The embedded tissues can be stored in wax for at least 2–3 years.
4. It is easier to form a ribbon during sectioning when the wax is at room temperature.
5. Longer probes give stronger signals. Usually, 1,000–1,500 bp would be a good size for a probe. If the gene of interest is smaller than 1,000 bp, you can use the entire coding region or include the untranslated regions (UTRs) for probes. One example of how a longer probe can improve signal strength is in 1. By using a longer *APETALA1* (*AP1*) probe, we were able to increase the signal strength and also distinguish *AP1* from its closely related gene (6).
6. The 3′ overhang can act as a promoter for RNA polymerases.
7. Alternatively, DNA template can be purified by phenol/chloroform extraction followed by ethanol precipitation.
8. Some transcription buffers contain spermidine, which will precipitate the template DNA at low temperature.
9. Alternatively, precipitate the probes by adding 10.5 μl of 4 M LiCl and 250 μl of cold 100% ethanol at −20°C for 20 min. Spin down the pellet and rinse it with 70% ethanol. Dry the pellet and resuspend it with 100 μl of DEPC-H_2O. However, probes tend to give higher backgrounds using this method.
10. RNA will not be well resolved in a non-denaturing gel. However, you can see multiple bright bands in the gel if the transcription reaction was successful.
11. A stir bar can be added to the staining dish to stir the solutions gently during these steps.

12. All solutions for the slide pretreatment should be prepared freshly with autoclaved ultrapure water on the hybridization day.

13. This step is critical for a successful in situ hybridization. Proteinase K partially digests tissues to allow better probe penetration. Insufficient digestion leads to high background while overdigestion causes tissue damage. The duration of proteinase K digestion might be different for other plant species. Always use freshly prepared proteinase K. Add proteinase K to the prewarmed solution to 1 µg/ml right before this step. Stir the solution to make sure that proteinase K is evenly distributed. Agitate the slides every 10 min during digestion.

14. Proteinase K digestion partially reverses fixation. Sections have to be fixed again by formaldehyde. Prepare the 3.7% formaldehyde in 1× PBS from the 37% formaldehyde solution right before use. Do not agitate the slides at this step.

15. This step acetylates positive charges in the sections to reduce the background (5). Add a stir bar to the staining dish. Dilute the 2 M stock of triethanolamine to 0.1 M. Let it stir for 30 s. Add acetic anhydride right before adding the slides. Stir gently using a stir bar throughout this step.

16. Sections have to be completely dried before proceeding to hybridization. Dried sections will appear white.

17. The hybridization buffer is very viscous and hard to pipet due to dextran sulfate. Always prepare excess amounts of buffer. After dextran sulfate is dissolved, do not put it on ice. Keep it either at room temperature or 65°C.

18. The optimal amount of probe varies among different genes. It is recommended to try several different amounts of probe, for example 1, 2, and 5 µl, when a gene is tested for the first time.

19. Prewarm the 0.2× SSC buffer in the hybridization incubator after finishing the hybridization. After the hybridization, slides have to be wet at all times or high background will develop.

20. Alternatively, ready-to-use alkaline phosphatase substrate solution, such as Western Blue, can replace NBT/BCIP here. 1 mM levamisole can be added to the substrate solution to reduce the background signal if this has become an issue.

21. 16–24 h of incubation is usually sufficient for strongly expressed genes. Replace with fresh substrate after two nights of incubation if the signal is still weak.

22. Save the ethanol series from the slide pretreatment. Start dehydration from 30% ethanol and end in histoclear before adding the mounting medium.

References

1. Pastore J et al (2011) LATE MERISTEM IDENTITY2 acts together with LEAFY to activate *APETALA1*. Development 138(15):3189–3198
2. Long JA, Barton MK (1998) The development of apical embryonic pattern in *Arabidopsis*. Development 125(16):3027–3035
3. Weigel D, Glazebrook J (2002) *Arabidopsis*: a laboratory manual. Cold Spring Harbor Laboratory, NewYork
4. Moench TR et al (1985) Efficiency of *in situ* hybridization as a function of probe size and fixation technique. J Virol Methods 11(2):119–130
5. Hayashi S et al (1978) Acetylation of chromosome squashes of *Drosophila* melanogaster decreases the background in autoradiographs from hybridization with [125I]-labeled RNA. J Histochem Cytochem 26(8):677–679
6. Blázquez M et al (2006) How floral meristems are built. Plant Mol Biol 60(6):855–870

Chapter 6

Laser Microdissection of Cells and Isolation of High-Quality RNA After Cryosectioning

Marta Barcala, Carmen Fenoll, and Carolina Escobar

Abstract

Laser capture microdissection (LCM) has become a powerful technique that allows analyzing gene expression in specific target cells from complex tissues. It is widely used in animal research, yet few studies on plants have been carried out. We have applied this technique to the plants–nematode interaction by isolating feeding cells (giant cells; GCs) immersed inside complex swelled root structures (galls) induced by root-knot nematodes. For this purpose, a protocol that combines good morphology preservation with RNA integrity maintenance was developed, and successfully applied to *Arabidopsis* and tomato galls. Specifically, early developing GCs at 3 and 7 days post infection (dpi) were analyzed; RNA from LCM GCs was amplified and used successfully for microarray assays.

Key words: Laser capture microdissection, RNA isolation, Cryosectioning, Arabidopsis, Tomato, Galls, Root-knot nematode, Giant cells

1. Introduction

Laser capture microdissection (LCM) is a technique that allows harvesting specific cells from complex tissues or populations for specific RNA, DNA, or protein isolation (1, 2) so that they can be used in downstream applications, such as microarray hybridization, cDNA library construction, proteomic analysis, etc. So far, several laser capture equipment have been developed by different companies (3). We performed LCM with the PixCell II system (Arcturus), which allows isolating cells using a low-power (infrared) laser and retaining them in a thermoplastic film; as the laser radiation is absorbed by the film instead of by the cell samples, this procedure should preserve the integrity of the captured material. The technique is applied to tissue sections that can be prepared from either frozen or paraffin-embedded biological samples. In general, better morphology can be observed in paraffin-embedded tissues fixed with non-coagulating fixatives, whereas the use of coagulating fixatives (such as ethanol: acetic acid) for frozen tissues renders higher RNA yield and quality (4). Thus, a compromise between both good morphology and RNA preservation should be achieved.

A few attempts to study specifically nematode-feeding sites (NFSs) have been carried out, reviewed in ref. 5. The first report was based on micro-aspiration of the cytosolic content of tomato giant cells (GCs) (6), but this method could only be applied to large GCs at late differentiation stages. In contrast, LCM represented an advance as it permits precise NFS collection at any infection time as long as they can be identified in histological sections. The first reported NFS isolation by LCM was applied to syncytia induced by cyst nematodes (7). We have developed a protocol to both preserve morphology and render high-quality RNA from GCs formed by root-knot nematodes at different developmental stages (3 and 7 days post infection (dpi)) of either *Arabidopsis* or tomato. Briefly, this protocol consists of a mild fixation step with a non-cross-linking fixative, gall cryosectioning, GCs LCM, and eventually RNA extraction. The RNA obtained by this protocol has been successfully used for microarray analysis.

Safety handling measures to avoid RNA degradation are strongly recommended, such as always wearing gloves, preparing all the solutions with diethyl pyrocarbonate (DEPC)-treated deionized water, and using RNase-free plastic and glassware. To prepare 0.1% DEPC-treated deionized water, add 1 ml DEPC to 1 L of deionized water (see Note 1), stir overnight at room temperature (RT), and autoclave it for 15 min at 100°C to destroy DEPC.

2. Materials

2.1. Tissue Fixation

1. Ethanol–acetic acid (EAA) fixative solution: Three parts of absolute ethanol (molecular biology grade) and one part of glacial acetic acid (3:1 v/v). Prepare it in a conical tube or a glass bottle by mixing the ethanol and acetic acid and keep it tightly closed on ice. Only freshly made fixative solution should be used.
2. Microcentrifuge tubes.
3. Surgical blade in a scalpel with handle.
4. Pointed tip tweezers.

2.2. Cryoprotective Solutions

1. 0.01 M phosphate-buffered saline (PBS) solution, pH 7.4: 0.138 M NaCl, 2.7 mM KCl. Dissolve a pouch in 1 L of DEPC-water (see Note 2) and store at RT.
2. 10% sucrose in 0.01 M PBS, pH 7.4. Add 5 g of sucrose to PBS for a final volume of 50 ml. Dissolve completely and store at 4°C (see Note 3).

3. 15% sucrose in 0.01 M PBS, pH 7.4. Add 7.5 g of sucrose to PBS for a final volume of 50 ml. Dissolve completely and store at 4°C.

4. 34.3% sucrose in 0.01 M PBS, pH 7.4. Add 17.25 g of sucrose to PBS for a final volume of 50 ml. Dissolve completely and store at 4°C.

5. 34.3% sucrose, 0.01% safranin-O dye, 0.01 M PBS, pH 7.4. Prepare this solution by adding safranin-O from a 1,000× stock to the previous solution.

6. 10% safranin-O (1,000×): Weigh 100 mg of safranin-O and add it to 1 ml of 0.01 M PBS, pH 7.4.

7. Vacuum Eppendorf concentrator 5301 (Eppendorf, Hamburg, Germany).

8. Orbital shaker.

2.3. Embedding and Cryosectioning

1. Embedding media: Tissue-TEK, optimal cutting temperature grade (O.C.T) media (Sakura Finetek, the Netherlands).
2. Cryomolds: Mold disposable base of 7×7×5 mm (see Note 4).
3. 2-Isopentane (methyl butane).
4. Liquid nitrogen.
5. Long forceps.
6. Poly-L-Lysine-coated slides: Polisyne (BDH, Poole, UK).
7. Vertical glass staining jars.
8. Ethanol solutions: 70 and 95% Ethanol.
9. Xylene.
10. Desiccant (silica gel).
11. Cryostat with disposable blades and anti-roll device. We have developed the protocol using a Leica CM3050S crysotat (Leica Microsystems, Wetzler, Germany).

2.4. Laser Capture Microdissection and RNA Extraction

1. PixCell II Laser Capture Microdissection system (Arcturus, Life Techonologies, California, USA).
2. CapSure HS LCM caps (Arcturus).
3. ExtracSure Sample Extraction device (Arcturus).
4. Absolutely RNA Nanoprep kit (Stratagene, California, USA).

3. Methods

3.1. Tissue Fixation

Unless otherwise specified, all the steps are carried out at 4°C by placing the microcentrifuge tubes on ice.

1. Clean the working surface and all metallic dissection instruments, such as the scalpel handle and the tweezers, with acetone.
2. Prepare aliquots with 1.5 ml of EAA fixative in as many microcentrifuge tubes as needed (see Note 5), and let them cool down on ice.
3. Localize the galls and root pieces to dissect under a stereo microscope; open the plate containing the in vitro-grown infected *Arabidopsis* plants and cover them with cool freshly made fixative EAA (see Note 6). Collect all the galls and control root pieces one by one, and transfer them into the fixative-filled tubes very quickly. Cut the samples carefully, leaving a small portion of root at both sides of the gall to facilitate sample handling with a pair of tweezers (i.e., when orienting the sample in the molds).
4. To facilitate fixative infiltration, apply vacuum to the samples for 15 min (see Note 7). Then, gently swirl the tubes for 1 h (i.e., in a shaker at 70 rpm).
5. Replace the solution with fresh fixative. Repeat the infiltration as in step 4 and swirl samples for 2 h (see Note 8).
6. Discard the fixative and add 10% sucrose in 0.01 M PBS, pH 7.4, to each tube. Apply vacuum for 10 min and swirl the samples for 3 h as described in step 4.
7. Continue with the cryoprotective treatment by infiltrating samples successively in 15% sucrose in 0.01 M PBS, pH 7.4, and 34.3% sucrose, 0.01% safranin-O, 0.01 M PBS, pH 7.4 (see Note 9) as described in step 6. Swirl the samples for 3 h and overnight, respectively, as described in step 4.
8. Rinse the samples in 34.3% sucrose in 0.01 M PBS, pH 7.4, to remove excess safranin-O dye immediately prior to embedding (see Note 10).

3.2. Embedding

1. Cool down isopentane almost to its freezing point by submersion in a liquid nitrogen bath (see Note 11) (Fig. 1a).
2. Label carefully the molds.
3. Fill a mold with Tissue Tek O.C.T. compound, taking special care to avoid air bubbles. This prevents molds from cracking during freezing and sectioning (see Note 12).
4. Carefully introduce a sample in the O.C.T. and orient it with the tweezers. It is important to drain as much sucrose solution as possible from the sample (see Note 13).
5. Freeze the sample by immersing the entire mold in precooled isopentane for at least 20 s. Large samples will need longer immersion times. Long, wide-open forceps are recommended as they facilitate handling of the mold during submersion into the isopentane without disturbing sample orientation (Fig. 1a). O.C.T. will turn white upon freezing.

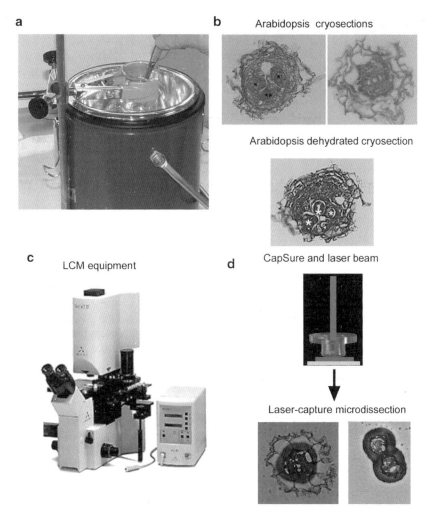

Fig. 1. Some steps of the LCM process. (**a**) Custom-made setup for cooling isopentane in a liquid nitrogen bath. (**b**) Cryosections of *Arabidopsis* galls before (*upper panels*) and after (*lower panel*) dehydration. Giant cells are labeled by white asterisks. (**c**) LCM PixCell II instrument. (**d**) The LCM process, showing a CapSure placed over a slide and the laser beam, an *Arabidopsis* root section after exposure to the laser beam, and the microdissected area captured onto a CapSure.

6. Transfer the mold to liquid nitrogen until you have finished freezing all the samples; then, store the molds at −80°C for later cryosectioning.

3.3. Cryosectioning and Laser Capture Microdissection

1. Let the cryostat cool down to −20°C (see Note 14). Then, place the samples inside and let them equilibrate to the cryostat temperature for at least 15 min. Also keep a staining jar filled with 70% ethanol inside the cryostat.

2. Trim off the mold and cut it in 10-µm sections. During cryosectioning, we check tissue morphology by picking a section in a slide (see the next step) and mounting it with glycerol to avoid desiccation. The section can then be observed and photographed under a microscope (Fig. 1b).

3. When GCs are discerned within a section, label a slide appropriately with a pencil and pick up the section. For this purpose, drop a slide (stored at room temperature) carefully onto the cut section/s. We usually mount 3–4 sections per slide (see Note 15). Keep, then, the slide inside the cryostat until you finish collecting the 3–4 sections, and then transfer it to the staining jar filled with 70% ethanol. Store the staining jar at −20°C until sectioning is completed (see Note 16).

4. Prior to LCM, the O.C.T. compound has to be removed and sections dehydrated (Fig. 1b). To do this, perform sequential washes in staining jars as follows: 70% ethanol (30 s, RT), 0.01 M PBS, pH 7.4 (30 s, RT), 70% ethanol (30 s, RT), 95% ethanol (−20°C, overnight), and 100% ethanol (30 s, RT). Finally, dehydrate sections completely by washing slides twice in xylene (10 min each) (see Note 17). Let slides air dry until xylene has completely evaporated, and store them at RT in a box with desiccant (such as silica gel). Sections should be used for LCM immediately.

5. Follow manufacturer's instructions for LCM (see Note 18; Fig. 1c).

6. Microdissect as many giant or vascular control cells per cap as possible. This number varies depending on the tissue size, position in the slide, etc. We usually collect around 50 giant cells (3 dpi) from either *Arabidopsis* or tomato galls, and 100 vascular control cells on each cap (see Note 19; Fig. 1d). Once the cap is full of microdissected tissue, attach the ExtracSure Sample Extraction device to the cap and store it at RT until finishing with LCM (see Note 20).

3.4. RNA Extraction

1. RNA extraction is performed with the Absolutely RNA Nanoprep kit. Manufacturer's instructions are followed with minor modifications (see Note 21).

2. Eluted RNA is stored at −80°C until subsequent amplification.

4. Notes

1. DEPC is highly toxic and volatile, and it must always be handled in a fume hood.

2. Instead of purchasing PBS buffer, it can be also prepared as described in ref. 8 and later it can be treated with DEPC.

3. Sucrose solutions become easily contaminated; therefore, we recommend preparing a reasonable volume and discard the remaining solutions once the procedure is completed.

4. Different base molds can be used according to the size of the processed tissue; usually, the smaller the base mold, the better, as large blocks tend to crack when being frozen.

5. In general, several samples can be fixed in each microcentrifuge tube, but this depends on the sample size, as enough fixative must be in contact and completely cover each sample. Usually, up to five 3-dpi galls were added per tube.

6. This step can be omitted when only a root piece has to be dissected from a plate. If several plants or root pieces have to be dissected from a single plate, it is advisable to cover the plate with approximately 5 ml of fixative to prevent root desiccation while the plate remains opened. Desiccation will collapse the tissue and destroy its morphology.

7. Depending on the tissue, vacuum treatment has to be empirically adjusted. Usually, 10–15 min is sufficient to make tomato or *Arabidopsis* root tissue sink to the bottom of the tube. Avoid strong vacuum pumps as the tissue could be damaged, and bubbling air can eject samples from the tube. To avoid this, tubes can be wrapped with parafilm-like plastic with some holes in it. It is important to point out that, when not using the same vacuum device as described in this protocol, the incubation times required for best results could vary considerably depending on the pump strength used.

8. Replace the fixative very carefully without damaging or losing the galls or the control roots. For that purpose, use a Pasteur pipette or a thin plastic tip and keep samples always floating in solution; do not let samples be uncovered at any time. It is recommended to leave less than 20 μl (ideally 10 μl) of the previous solution to avoid desiccation. It is also important to visually localize samples throughout the process in order to avoid having them get stuck to the tube walls.

9. The concentration of sucrose as cryoprotectant may be adjusted depending on the characteristics of the cells to be preserved. Additionally, although the use of dyes might not be advisable (9), there are exceptions with tissues (e.g., very young *Arabidopsis* galls or roots) that become nearly transparent and can be hardly identified inside the tubes, especially in later steps such as while embedding and cryosectioning.

10. As safranin-O is a water-soluble dye, long incubation in 0.01 M PBS solution will fade sample staining.

11. Work in a fume hood, as isopentane is very volatile. Fill half of a 100-ml beaker with isopentane and slowly transfer it to a Dewar flask containing liquid nitrogen. Leave the base of the beaker in contact with liquid nitrogen until solution becomes misty and thickens at the bottom. If a solid opaque white surface appears at the bottom (frozen isopentane), lift up the beaker out of the liquid nitrogen to warm it, and repeat the cooling process.

12. Keep the O.C.T. compound bottle upside down and squeeze it very gently when filling the cryomold to avoid bubble formation. If any bubble appears, remove it with a thin needle or

a pipette tip. Some other protocols recommend adding a thin layer of O.C.T. to the mold, then orient the sample, and finally fill the mold with O.C.T. In our hands, this was not adequate as we added several samples per mold (see Note 13), and the samples were easily disturbed.

13. Use very-thin-tip tweezers to take samples out of the tube and to orient them in the O.C.T. We usually place 2–3 galls or control roots per mold; this step is a bit tricky, mostly due to the bubbles that might be formed. Sucrose solution could be easily drained from the sample with a gentle touch onto a filter paper; however, this should be done with extreme care as it could overdry the sample. Therefore, we omitted this step during *Arabidopsis* gall processing.

14. Before cryosectioning, be sure that the anti-roll plate and the disposable blades have been cleaned and cooled down, since otherwise the sample could melt. It is also advisable to keep spare disposable blades inside the cryostat, just in case additional ones are needed.

15. At least for some laser capture microdissectors, technical limitations require sections to be centered in the slide. Therefore, when several sections per slide are needed, ideally a ribbon of 3–4 sections (as those from paraffin sectioning) are required; this implies becoming skilled in cryosectioning.

16. We usually place two 70% ethanol staining jars inside the cryostat. If there is not enough room inside the cryostat, prepare in advance several staining jars with 70% ethanol and keep them in the freezer.

17. Xylene is hazardous and volatile; always handle it in a fume hood.

18. We used an Arcturus PixCell II system for LCM. Laser beam power and length must be empirically established; for our tissue, it was typically set to 90–110 mW, 70 μs, and a spot size of 7.5 μm. Recommended materials from the manufacturer were used, such as Arcturus CapSure HS LCM caps for LCM and the ExtracSure sample extraction device for RNA extraction.

19. Sometimes, non-desired tissue remains attached to the cap after microdisecction. This can be easily eliminated by gently pressing the cap onto a post-it note without affecting RNA quality at all.

20. We processed caps immediately after LCM; it is not recommended to keep them at RT for more than 2 h, the longest time we tested with no detectable effects on RNA quality.

21. Caps were incubated in lysis buffer–β-mercaptoethanol mixture for 2–3 min at 60°C. DNase treatment was performed as indicated by the manufacturer. Finally, RNA was eluted twice

with 20 μl elution buffer pre-warmed to 60°C. For RNA extraction from tomato-microdissected GC, elution in a smaller volume (twice in 10 μl) has been successfully used.

Acknowledgments

The authors thank K. Lindsey and J. Topping (University of Durham, UK) for the use of their equipment and for their advice during the development of the microdissection protocol. Work in the laboratory was supported by grants from the Ministry of Education (AGL2007-60273) and the Ministry of Science and Innovation (AGL2010-17388) to CE, and PCI08-0074-0294 from the Junta de Comunidades de Castilla la Mancha and CSD2007-00057 from the Ministry of Science and Innovation to CF. M.B. was a recipient of a postdoc fellowship from the Consejería of Science and Education (JCMM) and the European Social Fund.

References

1. Bonner RF et al (1997) Laser capture microdissection: Molecular analysis of tissue. Science 278:1481–1483
2. Buck MR et al (1996) Laser capture microdissection. Science 274:998–1001
3. Bagnell C (2006) Laser capture microdissection. In: Coleman W, Tsongalis G (eds) Molecular diagnostics: for the clinical laboratorian. 2. Humana, Totowa, NJ, pp 219–224
4. Portillo M et al (2009) Isolation of rna from laser-capture-microdissected giant cells at early differentiation stages suitable for differential transcriptome analysis. Mol Plant Pathol 10:523–535
5. Escobar C, Sigal B, Mitchum MG (2011) Transcriptomic and proteomic analysis of the plant response to nematode infection. In: Fenoll JC, Gheysen G (eds) Genomics and molecular genetics of plant-nematode interactions, Springer Science, NY, pp 155–171
6. Wang Z, Potter R, Jones MGK (2001) A novel approach to extract and analyse cytoplasmic contents from individual giant cells in tomato roots induced by meloidogyne javanica. Int J Nematol 11:219–225
7. Ithal N et al (2007) Developmental transcript profiling of cyst nematode feeding cells in soybean roots. Mol Plant Microbe Interact 20:510–525
8. Sambrook J, Russell D (2001) Molecular cloning: a laboratory manual, vol 3, 3rd edn. Cold Spring Harbor Laboratory, Cold Spring Harbor, NY
9. Ramsay K, Wang Z, Jones M (2004) Using laser capture microdissection to study gene expression in early stages of giant cells induced by root-knot nematodes. Mol Plant Pathol 5:587–592

Chapter 7

Detection and Quantification of Alternative Splicing Variants Using RNA-seq

Douglas W. Bryant Jr, Henry D. Priest, and Todd C. Mockler

Abstract

Next-generation sequencing has enabled genome-wide studies of alternative pre-mRNA splicing, allowing for empirical determination, characterization, and quantification of the expressed RNAs in a sample in toto. As a result, RNA sequencing (RNA-seq) has shown tremendous power to drive biological discoveries. At the same time, RNA-seq has created novel challenges that necessitate the development of increasingly sophisticated computational approaches and bioinformatic tools. In addition to the analysis of massive datasets, these tools also need to facilitate questions and analytical approaches driven by such rich data. HTS and RNA-seq are still in a stage of very rapid evolution and are, therefore, only introduced in general terms. This chapter mainly focuses on the methods for discovery, detection, and quantification of alternatively spliced transcript variants.

Key words: RNA-seq, Bioinformatics, Next-generation sequencing, Transcript abundance, Alternative splicing

1. Introduction

Over the past several years, the advent of multiple competing and complementary massively parallel DNA sequencing platforms, the so-called next-generation sequencing (NGS) revolution, has completely transformed the landscape of molecular genetic studies, enabling many new biological applications and techniques. One such newly enabled application, known as high-throughput RNA sequencing (RNA-seq), makes possible genome-wide studies of alternative pre-mRNA splicing. Leveraging the unprecedented quantity of data produced by HTS platforms along with the exquisite sensitivity of RNA-seq experiments enables empirical determination, characterization, and quantification of the expressed RNAs in a sample in toto. Levels of individual transcripts, alternative splicing, other transcript processing events, and allele-specific gene expression can be simultaneously surveyed using RNA-seq. Further, RNA-seq facilitates the empirical annotation of transcribed genome regions, including transcript splice variants within and across samples. While RNA-seq has shown tremendous power to drive biological discoveries, it has also delivered challenges that have necessitated the development of increasingly sophisticated computational approaches and bioinformatic tools, not only to deal with massive datasets but also to facilitate questions and analytical approaches driven by such rich data. Because HTS and RNA-seq are in a stage of very rapid evolution, this chapter mainly focuses broadly on the methods for discovery, detection, and quantification of alternative splicing transcript variants.

2. Materials and Methods

2.1. RNA Isolation and Ribo-Depletion

RNA-seq experiments all begin with isolation of RNA from the samples of interest and subsequent RNA-seq library preparation. Discussion of RNA isolation in detail is beyond the scope of this chapter as it is a broad area with numerous protocols and considerations depending on the samples involved, such as tissue type and species involved. It is important to note that while strictly speaking the term "transcriptome" refers broadly to the entire complement of transcribed sequences in a genome, here we are referring specifically to RNA-seq conducted on the mRNA component, which is the transcript species of interest for most studies of alternative splicing. In these studies, it is most often desirable to focus on the mRNA component by thoroughly depleting the sample of non-mRNA species, especially the large proportion (>90%) of ribosomal RNA (rRNA) in isolated total RNA. Thus, the depletion of the abundant rRNA species is a critical first step in most RNA-seq experiments. Depletion of rRNA can be achieved in several ways. One of the most common approaches is isolation of polyadenylated (polyA+) mRNAs by oligo-dT affinity purification using any one of a number of commercially available kits. However, the resulting polyA + mRNA fraction is not only depleted in rRNAs, but also depleted in several non-polyA RNA species, including non-polyA mRNAs, tRNAs, and other transcripts of unknown function (reviewed in ref. 1). Alternative approaches for ribo-depletion that retain the potentially interesting non-polyA transcripts exist. Probably, the most widely used approach involves selective depletion of rRNA molecules from total RNA samples by hybridization to rRNA sequence-specific biotin-labeled locked nucleic acids (LNAs) (e.g., Invitrogen RiboMinus™ technology), followed by their removal with streptavidin-coated magnetic beads. Regardless of the exact method employed for ribo-depletion, avoiding degradation of the mRNA prior to RNA-seq library preparation is essential.

2.2. Library Preparation and RNA-seq

The next steps in RNA-seq greatly depend on the nature of the samples being examined, the quantity of available mRNA after ribo-depletion, the type of RNA-seq data desired (whether single end or paired end, nondirectional or directional), and the HTS platform that is used to generate the data. The first generation of RNA-seq protocols was based on adaptations of the first next-gen sequencing protocols developed for HTS of genomic DNA. Generally, these approaches involved sequencing of a double-strand cDNA library prepared by fragmenting double-stranded cDNAs generated by reverse transcription using either random hexamer or oligo-dT primers (e.g., refs. 2–5). The significant disadvantage of these approaches, in particular for oligo-dT-primed cDNAs, is the tendency of reverse transcriptase to disassociate from the template, resulting in pronounced 3′ bias, and which may lead to underrepresentation of sequence data for the 5′ ends of transcripts. The use of random primers alleviated this problem (5) by providing much better representation of the 5′ ends of transcripts. In summary, these approaches relied on the initial generation of high-quality, full-length, enriched or random-primed cDNAs in order to capture the entire transcript structures, but suffered from a total loss of directional information due to the way the sequencing adaptors were ligated to the cDNA fragments and subsequent amplification steps.

Alternative RNA-seq methods involving fragmentation of RNA instead of cDNA have proliferated because they offer several advantages. First, by utilizing relatively small fragments of RNA for library preparation, artifacts induced by possible secondary structures can be reduced, yielding more uniform coverage of reads over the lengths of transcripts. In the simplest embodiments of these methods, the RNA is first fragmented using thermal, chemical, or enzymatic hydrolysis. Next, the fragmented RNA is ligated to adapters. Finally, the fragments are reverse transcribed using primers complementary to the adapters. Several elaborations of the RNA fragmentation approach have been developed (6–8) to facilitate strand-specific sequencing.

Current state-of-the-art approaches for RNA-seq library preparation aim to preserve transcript directionality, which can be instrumental for downstream data analysis of alternative splicing. Knowing the strand of origin for RNA-seq data enables direct inference of gene orientation in the genome, allows for detection of overlapping transcripts, and ensures that overlapping transcripts from different gene loci are not erroneously interpreted as being splice variants of a single gene. To address this important issue, a number of approaches have been taken. Some protocols involve treating RNA with sodium bisulfite to convert cytosine residues to uracil (9); other methods involve strand-specific ligation of RNA adapters to the RNA template (10–12); while others insert strand-specific adaptor sequences into double-stranded cDNAs during their synthesis; yet another method involves incorporation of dUTP during synthesis of second-strand cDNA and subsequent selective digestion of the dUTP-containing strand with uracil-*N*-glycosylase enzyme (7). Several kits for strand-specific RNA-seq library preparation are commercially available from vendors, including Illumina (http://www.illumina.com/), Epicentre Biotechnologies (http://www.epibio.com), Invitrogen (http://www.invitrogen.com), Nugen (http://www.nugeninc.com), and others, and as an area of active development new protocols and commercial kits are released frequently.

The next step in RNA-seq protocols involves amplification of cDNA fragments prior to sequencing on the HTS platform of choice. In the case of Illumina, fragments bearing adaptor sequences are deposited on an oligonucleotide primer-functionalized glass flow cell surface, and solid-phase amplification ("cluster generation") is performed to produce randomly distributed, spatially separated template clusters at high density. The amplified clusters contain adaptor sequences needed to initiate the Illumina sequencing-by-synthesis reactions. In the SOLiD and 454/Roche systems, clonally amplified templates are prepared by emulsion PCR (emPCR). In emPCR, cDNA fragments are denatured into single-stranded fragments and captured on beads with a target stoichiometry of one fragment–one bead, and then amplified in situ. After amplification, the beads are immobilized on a glass surface (SOLiD) by chemical cross-linking or deposited into individual wells (454), after which the sequencing chemistry is performed on the respective HTS platform.

2.3. Short Read Assembly

From the HTS platform, short reads, representing the sample's expressed transcripts, are obtained. These reads can number in millions, or even billions, depending on the chosen HTS platform and experimental scale. Regardless of downstream application, these reads first need to be either mapped to a genomic reference or assembled de novo.

2.3.1. Mapping to Reference

For mapping RNA-seq reads to a genomic reference, necessary components include:

1. RNA-seq short reads.
2. Genomic reference.
3. Short-read aligner.

Mapping RNA-seq data represents a special case of the well-studied general read-mapping problem. While many approaches exist for the general case of mapping reads, the special case of mapping RNA-seq data is more challenging for several reasons. First, the reads generated by the highest-throughput platforms (Illumina and SOLiD) are typically very short, on the order of 30–150 base pairs. Second, per-base sequencing error rates, which tend to be most pronounced near the 3′ ends of reads, can be considerable, decreasing mappability or increasing the uncertainty of mapping. Third, the scale of the alignment problem can be daunting when dealing with hundreds of millions of reads from a single experiment.

Each read generated in an RNA-seq experiment either originates from a single exon, without crossing an exon–exon boundary, or originates from a pair of exons, spanning an intron. These two cases, unspliced alignments and spliced alignments, respectively, require different mapping approaches.

We first examine a stepwise, wholly empirical method using Burrows–Wheeler transform algorithms (BWA) (13), supersplat (14), and gumby (15) after which we examine a method using TOPHAT (16).

2.3.2. Gumby: Unspliced Alignments

From the total set of RNA-seq reads, those reads which fully align to the reference, unspliced reads, are identified first. Any modern read alignment application may be used in this first step, and indeed many such applications exist. Seed-based alignment algorithms (e.g., SHRiMP, BFAST, SeqMap, SOAP, MAQ, STAMPY) identify matches to short K-mer subsequences (the "seeds") within reads and then attempt to extend the seed matches to full alignments (17–24). The Burrows-Wheeler transform algorithms (BWA, SOAP2, Bowtie) use a data compression technique to compress the genome into a format that is optimized for searching for perfect matches but accommodates mismatches with a modest performance hit (13, 25, 26). The hash-based aligners (e.g., HashMap (5)) identify matches to an index of k-mers derived from the exact reference genome or transcriptome, where k equals the read length. This approach performs alignments very rapidly, but is limited to perfect match alignments and requires much more physical random access memory (RAM) than other alignment algorithms.

Here, we focus on BWA, which consists of three sub-applications executed in series. To align RNA-seq data to a genomic reference, the genomic reference must first be indexed:

```
% bwa index [reference.fasta]
```

Here, [reference.fasta] represents the name of the genomic reference file. Depending on the genomic reference's size, this indexing process can take up to 1 or 2 h, resulting in several files being created within the reference.fasta directory.

Next, the table of contents SAI file is created:

```
% bwa aln -t [NumThreads] [reference.fasta]
  [reads.fastq] > [reads.sai]
```

Here, "–t [NumThreads]" allows bwa aln to utilize more than one thread if desired in order to reduce runtimes. [reference.fasta] represents the same genomic reference file indexed in the first step, and [reads.fastq] represents the set of RNA-seq reads which can be in FASTA or FASTQ format. Finally, [reads.sai] represents the name of a new file to be created by BWA in this step.

In the third and final step, alignments are generated. For single-end reads, typical in RNA-seq experiments:

```
% bwa samse [reference.fasta] [reads.sai] [reads.
  fastq] -f [out.sam]
```

Here, [reference.fasta] again represents the same genomic reference file indexed in the first step. [reads.sai] represents the output from the second step, and [reads.fastq] again represents the set of RNA-seq reads. Finally, [out.sam] represents the name of the SAM-formatted alignment file to be produced. Depending on the size of the input reads set, this alignment step can take several hours.

2.3.3. Gumby: Spliced Alignments

As output in the final unspliced alignments step, out.sam contains both all alignments identified by BWA and all reads which BWA failed to align. These unaligned reads may be identified by the out.sam file's second column, which contains the numeral 4. Generally, a simple script is used to separate aligned reads from unaligned reads, creating a SAM-formatted file for each case. The set of aligned reads is set aside while the unaligned reads are processed.

Each unaligned read may have failed alignment due to containing errors, having low complexity, or being truly spliced, spanning an intron. To separate low-complexity reads from true spliced reads, typically a filter is applied, such as DUST (NCBI C Toolkit; http://www.ncbi.nlm.nih.gov/IEB/ToolBox/index.cgi). To remove low-complexity reads with DUST, first, the set of unaligned reads needs to be output in FASTA format with a simple script. Then, DUST is executed as:

```
% dust [unaligned_reads.fasta] > [unaligned_fil-
  tered_reads.fasta]
```

Here, [unaligned_reads.fasta] represents the FASTA-formatted file containing all unaligned reads and [unaligned_filtered_reads.fasta] represents the newly created file which contains masked reads, with low-complexity regions masked with "N"s. Finally, a simple script is used to remove all N-containing reads from [unaligned_filtered_reads.fasta], leaving only reads of sufficient complexity for the next step.

To identify putative splice junctions, all unaligned reads which passed the complexity filter are input to supersplat, which identifies all locations in the genomic reference, where reads can be spliced-aligned, guided by minimum and maximum intron size and minimum flank length. Supersplat is run in two steps. First, splat is executed with the following parameters:

```
%  supersplat    splat    -r    [reference.fasta]
   -f [filtered_reads.fasta]    -c    [MinChunkSize]
   -n [MinIntronSize]        -x    [MaxIntronSize]
   -i [MaxIndexChunkSize]    -t    [NumThreads]
   -N [MinReadCopyNumber]    -e    [OutputFileName]
```

Here,

- [reference.fasta] represents the same genomic reference file used to align reads with BWA.
- [filtered_reads.fasta] represents the unaligned and filtered RNA-seq reads.
- [MinChunkSize] represents the shortest read chunk size, typically 6–8.
- [MinIntronSize] represents the length of the smallest intron identified.
- [MaxIntronSize] represents the length of the longest intron identified.
- [NumThreads] represents the number of threads splat should use.
- [MinReadCopyNumber] restricts splat alignments to reads of this many copies or greater.
- [OutputFileName] represents the final output alignment file.

Output from this step is an alignment file containing all putative intron locations supported by the evidence. Since this set of alignments is exhaustive containing every location where an input read can be spliced-aligned within the set of used parameters, it is necessary to filter these results. This filtering is accomplished with supersplat stack:

```
% supersplat    stack    -r    [reference.fasta]
  -s [splat.output]    -n    [MinUniqueHits]
  -c [CopyNumber]      -o    [OutputFileName]
```

Here,

- [reference.fasta] represents the same genomic reference file used to align reads with BWA.
- [splat.output] represents the output from splat, above.
- [MinUniqueHits] represents the minimum number of unique reads mapped to an intron for that intron to pass the filter.
- [CopyNumber] acts the same way it does in splat, representing the minimum number of read copies for the read to pass the filter.
- [OutputFileName] represents the stack output file name.

Successfully aligned reads output by BWA along with the set of spliced-aligned reads output by supersplat are now used as input to the empirical gene-building application gumby:

```
% gmb -r [reference.fasta] -a [alignments.sam]
  -s [supersplat.out] -t [NumThreads] -o [output.gff3]
```

- Here, [reference.fasta] again represents the same genomic reference file used to align reads with BWA, [alignments.sam] represents BWA's aligned reads in SAM format, [supersplat.out] represents the output obtained in the last step from supersplat, [NumThreads] represents the number of threads gumby should use, and [output.gff3] represents the final GFF3-formatted output. This final GFF3-formatted output represents all gene models present in and identified from the original input set of RNA-seq reads. This output can be visualized in an application, such as GBrowse, or used for any other downstream analysis.

2.3.4. TopHat: Alignments

Several tools exist which perform both the exonic and spliced mappings, such as TopHat (16), MapSplice (27), and SpliceMap (28). Here, we focus on TopHat.

TopHat identifies putative splice junctions by mapping RNA-seq reads to the reference genome with the associated short-read alignment application BowTie (13). TopHat then builds a database of possible splice junctions, followed by mapping the previously unmapped reads against this junction database.

To use TopHat, several applications must be present in your UNIX PATH, including:

1. Bowtie—short-read aligner.
2. Bowtie-inspect—application which queries bowtie indexes to determine their type.
3. Bowtie-build—application which builds a bowtie index from DNA sequences.
4. Samtools—a set of applications for manipulating SAM- and BAM-formatted files.

The genomic reference is first prepared by creating a bowtie index:

```
% bowtie-build [reference.fasta][ReferenceName]
```

Here, [reference.fasta] represents the FASTA-formatted genomic reference file and [ReferenceName] represents a name given to the reference.

Next, tophat is executed:

```
% tophat [PathToIndex] [reads.fastq]
```

Here, [PathToIndex] represents the name assigned to the index in the previous step, and [reads.fastq] the set of RNA-seq reads to be aligned.

Several files are output, including:

- accepted_hits.bam—list of SAM-formatted read alignments.
- junctions.bed—BED-formatted list of splice junctions.

- insertions.bed—BED-formatted list of insertions.
- deletions.bed—BED-formatted list of deletions.

These output files may be imported directly to a number of genome viewers or may be used for other downstream analyses.

2.3.5. De Novo Assembly Without a reference genome or transcriptome, it is impossible to apply the read-mapping approach described in the previous section for the analysis of RNA-seq data. In such cases, de novo assembly of RNA-seq reads to generate full-length draft transcript sequences is required (reviewed in refs. 29–32). Although a detailed discussion is beyond the scope of this chapter, a worthwhile preliminary step prior to de novo assembly is to perform quality assessment, quality filtering, read trimming, and/or error correction (33–36) prior to assembly, any or all of which is likely to both improve the quality of the resulting transcript sequences and reduce the computational demands during assembly. The current state of the art in de novo assembly of RNA-seq data involves using tools, such as Velvet (37), Abyss (38), transAbyss (39), or Trinity (40), that use a de Bruijn graph (41) structure to resolve transcript sequence assemblies (reviewed in refs. 30, 31). As with any short-read assembly challenge, strategies for improving the quality of the assembly include acquiring more read data, combining data from different platforms (e.g., Illumina and 454), using paired-end reads, and using reads derived from different source libraries. Assembly algorithms commonly struggle to distinguish sequencing error from genuine sequence variation (SNPs), and assembly results are sensitive to various algorithm parameters (e.g., k-mer length, coverage cutoffs). De novo assembly of RNA-seq data presents some particular challenges because the abundance of transcripts can span orders of magnitude, with some genes represented by zero or very few transcripts, whereas other genes are represented by thousands of transcripts. This is a particular problem for de novo approaches that utilize k-mer coverage cutoffs to follow or discard paths through the graph structure. Some algorithms have implemented strategies to deal with this problem. For example, transABySS (39) uses variable k-mer lengths that facilitate transcript assembly over a wide range of expression levels. Other difficulties arise from alternative splicing such that reads representing the same gene but arising from distinct transcript isoforms can be misassembled, leading to a collapse of multiple isoforms to one or few putative transcript isoforms. Trinity attempts to address this problem by using a dynamic programming procedure to identify paths through the graph that are supported by actual reads to resolve ambiguities and reduce the number of potential paths to those representing actual transcript structures (40). RNA-seq reads can also be derived from unspliced precursor

2.4. Inference and Quantification of Alternative Transcript Isoforms

RNAs or partially processed transcripts, making it challenging to distinguish mature processed transcripts from noise. As with any other de novo assembly strategy—for example genome assembly—there are challenges not specific only to RNA-seq data.

A significant advantage of RNA-seq compared to other technologies, such as DNA microarrays, is its theoretically straightforward ability to define and analyze expressed transcripts at the isoform level. As discussed in previous sections, this entails first detecting and/or assembling alternative transcript isoforms, and then quantifying different isoforms in terms of their transcript abundance. Simply detecting or assembling distinct transcript variants does not provide a quantitative estimate of their relative abundances within a single sample or their differential expression between samples (e.g., tissues, conditions, etc.). Moreover, estimating isoform-specific expression is difficult because many genes have multiple isoforms or close paralogs. For such genes, it is often difficult, if not impossible, to assign some reads to a particular isoform or paralog.

Whether focused on previously known (e.g., annotated) splice variants or potentially novel splice variants predicted de novo from RNA-seq data, the most conservative approach involves quantification of transcript isoforms using only those sequences which are unique to particular isoforms (5). In this approach, the sequences of the known or predicted transcript variants for a given gene are compared a priori in order to define a set of "informative" sequences that can specifically differentiate each variant from the others. The informative sequences may represent gene features unique to a particular isoform, such as introns, exons, or splice junctions. The presence of RNA-seq reads mapping to such informative sequences confirms the expression of the corresponding splice variants unambiguously, and the normalized counts of reads mapping to these informative sequence features can be used to estimate relative abundances of such transcript variants. Similarly, the Alexa-seq method (42) estimates isoform-specific expression using only those reads mapping uniquely to one isoform. Such approaches fail for isoforms that do not contain unique exons or that contain very minor sequence differences (e.g., slight variation in donor/acceptor sites) that will be reflected in few RNA-seq reads. Moreover, these methods also directly depend on and require prior knowledge of the precise annotation or transcript structures of splice variants.

Alternative approaches include those that attempt to infer the relative abundances of transcript variants by considering all RNA-seq data mapping to a gene, thereby circumventing the problem of uncertainty in read assignments (reviewed in refs. 30, 31). The algorithms that implement this approach include Cufflinks (43) and MISO (44). Generically speaking, these algorithms attempt to mitigate uncertainty in read assignments to particular isoforms by modeling the RNA-seq process and estimating isoform abundances that best fit the observed RNA-seq data (39, 43–48). Results are, thus, an estimate of isoform abundance, but are sensitive to the overall gene expression level, the number of isoforms under consideration, and uncertainties introduced by misassembly. In particular, the maximum likelihood estimates for low-abundance transcripts can be inaccurate.

A common aim is to quantify the differences in the abundances of transcript isoforms over multiple samples using RNA-seq. As described in the previous section, various methods exist for quantifying the expression of alternatively spliced transcript isoforms. Assuming that the transcript structures have been assembled and quantified, there are several approaches and tools for detecting differential expression of isoforms between samples. The Cufflinks (43) pipeline both infers transcript isoforms directly from RNA-seq reads and performs statistical analysis to assess differential expression of the inferred transcripts. In cases where a researcher has inferred transcript isoforms and estimated their expression levels with other approaches, there are a variety of tools available for applying statistical tests for analysis of differential isoform expression. These methods (reviewed in refs. 30, 31) typically attempt to estimate the significance of differences between samples with few biological replicates. Examples include EdgeR (49), DESeq (50), Myrna (51), and GENE-counter (52), which attempt to model variance in read counts over replicates using various approaches, including negative binomial (NB) distribution (49, 50, 52).

3. Challenges and Future Directions

Sequencing technology will continue advancing and evolving unabated. The recent introduction of the so-called personal genome sequencing (PGS) platforms (e.g., Ion Torrent and Illumina MiSeq) will further accelerate the ability to generate transcriptome data, leading to the routine use of RNA-seq to analyze alternative splicing in the near future. Even though significant achievements have been made in terms of elucidating details of alternative splicing using RNA-seq, we are far from having a comprehensive view and understanding of alternative transcript processing and its biological consequences in any species.

Even without considering the advantages and challenges imposed by improved HTS technologies per se, there are a number of improvements needed for RNA-seq experimental methodologies and computational tools for analysis of alternative splicing using RNA-seq data. RNA-seq protocols are typically more complex than DNA-seq protocols and extremely sensitive to challenges, such as contamination or degradation of RNA. Most methods require greater quantities of total isolated or polyA-enriched/ribo-depleted RNA than can be generated in many experiments involving limiting samples. In cases such as these, there are methods for amplifying libraries, but such amplification efforts can introduce biases and artifacts. Strand-specific methods need to be more widely adopted in order to facilitate correct determination of gene directionality and ensure that overlapping transcripts representing different genes in the same region are not incorrectly identified as splice variants of the same gene. Paired-end methods need to be widely used to ensure that when multiple alternative splicing events affect transcripts from the same gene they are correctly identified as either co-occurring in the same transcript molecule or occurring in separate molecules. The difficulty and expenses of RNA-seq library production methods coupled with the relatively high cost of HTS have also presented another problem—the lack of biological replication in most RNA-seq-based studies of alternative splicing published to date. Improved library methods coupled with multiplexing on HTS platforms with ever-increasing throughput—and hence ever-decreasing costs per base sequenced—should drive greater use of biological replication in splicing studies, which will allow more precise quantification of each transcript isoform, facilitate detection of differential expression of splice variants, and enable inferences about real functional and biologically relevant differences between samples (e.g., individuals, strains, tissues, cell types, treatments).

While HTS-driven RNA-seq studies have been delivering a revolution in our ability to study transcriptomes, the disconnect between the technical capability to generate data and our ability to mine those data to elucidate biological discoveries has been exacerbated. There needs to be continuous development of algorithms, computational strategies, and human talent to handle massive RNA-seq datasets. As HTS platforms proliferate and the underlying sequencing technologies change or mature, bioinformatic solutions will need to evolve. For example, recently, we have seen dramatic increases in typical read lengths in RNA-seq data. We can reasonably expect this trend to continue, which will require new algorithms and tools to accurately and efficiently map millions and billions of reads to genomes. On the other hand, characterization of splice variants will benefit from longer reads that span multiple splice junctions and help resolve the "trans-frag" problem, so the mapping challenges may be viewed as a welcome trade-off. Other computational challenges that need to be addressed include improved methods for transcript annotation, and transcriptome reconstruction from RNA-seq data. Current algorithms can predict transcript structures but frequently predict far more transcript models than can be reasonably expected for a given alternatively spliced locus. So we need better approaches for differentiating true-positive versus false-positive transcript variants, which will again benefit from greater use of longer, strand-specific, and paired-end RNA-seq data. New approaches for estimating expression of individual transcript variants from RNA-seq data need to be developed, including new statistical approaches and wider use of true biologically replicated datasets.

In summary, RNA-seq is a powerful tool for characterizing and quantifying alternative splicing. Like any new technology or experimental platform that is still in its relative infancy, there are technical and analytical challenges remaining to be solved. The continuing improvements to methods, sequencing platforms, and computational solutions, coupled with continued decreasing costs, will expand the application of RNA-seq as an experimental tool for studying alternative splicing, and will thus lead to basic discoveries about the mechanisms and functions of this important system for increasing transcript diversity in vivo.

References

1. Jacquier A (2009) The complex eukaryotic transcriptome: unexpected pervasive transcription and novel small RNAs. Nat Rev Genet 10:833–844
2. Fox S, Filichkin S, Mockler TC (2009) Applications of ultra-high-throughput sequencing. Methods Mol Biol 553:79–108
3. Mortazavi A et al (2008) Mapping and quantifying mammalian transcriptomes by RNA-Seq. Nat Methods 5:621–628
4. Nagalakshmi U et al (2008) The transcriptional landscape of the yeast genome defined by RNA sequencing. Science 320: 1344–1349
5. Filichkin SA et al (2010) Genome-wide mapping of alternative splicing in Arabidopsis thaliana. Genome Res 20:45–58
6. Li H et al (2008) Determination of tag density required for digital transcriptome analysis: application to an androgen-sensitive prostate cancer model. Proc Natl Acad Sci USA 105:20179–20184
7. Parkhomchuk D et al (2009) Transcriptome analysis by strand-specific sequencing of complementary DNA. Nucleic Acids Res 37:e123
8. Ingolia NT et al (2009) Genome-wide analysis in vivo of translation with nucleotide resolution using ribosome profiling. Science 324:218–223
9. He Y et al (2008) The antisense transcriptomes of human cells. Science 322:1855–1857
10. Cloonan N et al (2008) Stem cell transcriptome profiling via massive-scale mRNA sequencing. Nat Methods 5:613–619
11. Lister R et al (2008) Highly integrated single-base resolution maps of the epigenome in Arabidopsis. Cell 133:523–536
12. Core LJ, Waterfall JJ, Lis JT (2008) Nascent RNA sequencing reveals widespread pausing and divergent initiation at human promoters. Science 322:1845–1848
13. Li R et al (2009) SOAP2: an improved ultra-fast tool for short read alignment. Bioinformatics 25:1966–1967
14. Bryant DW et al (2010) Supersplat-spliced RNA-seq alignment. Bioinformatics 26:1500–1505
15. Bryant DW, et al (2011) Gumby—a purely empirical RNA-seq-based approach to genome annotation. Manuscript in Preparation.
16. Trapnell C, Pachter L, Salzberg SL (2009) TopHat: discovering splice junctions with RNA-Seq. Bioinformatics 25:1105–1111
17. Jiang H, Wong WH (2008) SeqMap: mapping massive amount of oligonucleotides to the genome. Bioinformatics 24:2395–2396
18. Li H, Ruan J, Durbin R (2008) Mapping short DNA sequencing reads and calling variants using mapping quality scores. Genome Res 18:1851–1858
19. Li R et al (2008) SOAP: short oligonucleotide alignment program. Bioinformatics 24: 713–714
20. Smith AD, Xuan Z, Zhang MQ (2008) Using quality scores and longer reads improves accuracy of Solexa read mapping. BMC Bioinformatics 9:128
21. Homer N, Merriman B, Nelson SF (2009) BFAST: an alignment tool for large scale genome resequencing. PLoS One 4:e7767
22. Rumble SM et al (2009) SHRiMP: accurate mapping of short color-space reads. PLoS Comput Biol 5:e1000386

23. Lunter G, Goodson M (2011) Stampy: a statistical algorithm for sensitive and fast mapping of Illumina sequence reads. Genome Res 21:936–939
24. Rizk G, Lavenier D (2010) GASSST: global alignment short sequence search tool. Bioinformatics 26:2534–2540
25. Li H et al (2009) The sequence alignment/map format and SAMtools. Bioinformatics 25:2078–2079
26. Li H, Durbin R (2009) Fast and accurate short read alignment with Burrows-Wheeler Transform. Bioinformatics 25:1754–1760
27. Wang K et al (2010) MapSplice: accurate mapping of RNA-seq reads for splice junction discovery. Nucleic Acids Res 38:e178
28. Au KF et al (2010) Detection of splice junctions from paired-end RNA-seq data by SpliceMap. Nucleic Acids Res 38:4570–4578
29. Denoeud F et al (2008) Annotating genomes with massive-scale RNA sequencing. Genome Biol 9:R175
30. Garber M et al (2011) Computational methods for transcriptome annotation and quantification using RNA-seq. Nat Methods 8:469–477
31. Costa V et al (2010) Uncovering the complexity of transcriptomes with RNA-Seq. J Biomed Biotechnol 2010:853916
32. Yassour M et al (2009) Ab initio construction of a eukaryotic transcriptome by massively parallel mRNA sequencing. Proc Natl Acad Sci USA 106:3264–3269
33. Kelley DR, Schatz MC, Salzberg SL (2010) Quake: quality-aware detection and correction of sequencing errors. Genome Biol 11:R116
34. Shi H et al (2010) A parallel algorithm for error correction in high-throughput short-read data on CUDA-enabled graphics hardware. J Comput Biol 17:603–615
35. Yang X, Dorman KS, Aluru S (2010) Reptile: representative tiling for short read error correction. Bioinformatics 26:2526–2533
36. Kao WC, Chan AH, Song YS (2011) ECHO: a reference-free short-read error correction algorithm. Genome Res 21:1181–1192
37. Zerbino DR, Birney E (2008) Velvet: algorithms for de novo short read assembly using de Bruijn graphs. Genome Res 18:821–829
38. Birol I, Jackman SD, Nielsen CB (2009) De novo transcriptome assembly with ABySS. Bioinformatics 25:2872–2877
39. Robertson G et al (2010) De novo assembly and analysis of RNA-seq data. Nat Methods 7:909–912
40. Grabherr MG et al (2011) Full-length transcriptome assembly from RNA-Seq data without a reference genome. Nat Biotechnol 29:644–652
41. De Bruijn NG (1946) A combinatorial problem. Koninklijke Nederlandse Akademie v Wetenschappen 46:6
42. Griffith M et al (2010) Alternative expression analysis by RNA sequencing. Nat Methods 7:843–847
43. Trapnell C et al (2010) Transcript assembly and quantification by RNA-Seq reveals unannotated transcripts and isoform switching during cell differentiation. Nat Biotechnol 28:511–515
44. Katz Y et al (2010) Analysis and design of RNA sequencing experiments for identifying isoform regulation. Nat Methods 7:1009–1015
45. Marioni JC et al (2008) RNA-Seq: an assessment of technical reproducibility and comparison with gene expression arrays. Genome Res 18:1509–1517
46. Jiang H, Wong WH (2009) Statistical inferences for isoform expression in RNA-Seq. Bioinformatics 25:1026–1032
47. Li B et al (2009) RNA-Seq gene expression estimation with read mapping uncertainty. Bioinformatics 26:493–500
48. Richard H et al (2010) Prediction of alternative isoforms from exon expression levels in RNA-Seq experiments. Nucleic Acids Res 38:e112
49. Robinson MD, McCarthy DJ, Smyth GK (2010) edgeR: a bioconductor package for differential expression analysis of digital gene expression data. Bioinformatics 26:139–140
50. Anders S, Huber W (2010) Differential expression analysis for sequence count data. Genome Biol 11:R106
51. Langmead B, Hansen KD, Leek JT (2010) Cloud-scale RNA-sequencing differential expression analysis with Myrna. Genome Biol 11:R83
52. Cumbie JS, et al (2011) GENE-counter: a computational pipeline for the analysis of RNA-Seq data for gene expression differences. PLoS One (6):e25279

Chapter 8

Separating and Analyzing Sulfur-Containing RNAs with Organomercury Gels

Elisa Biondi and Donald H. Burke

Abstract

Polyacrylamide gel electrophoresis is a widely used technique for RNA analysis and purification. The polyacrylamide matrix is highly versatile for chemical derivitization, enabling facile exploitation of thio-mercury chemistry without the need of tedious manipulations and/or expensive coupling reagents, which often give low yields and side products. Here, we describe the use of [(N-acryloylamino)phenyl]mercuric chloride in three-layered polyacrylamide gels to detect, separate, quantify, and analyze sulfur-containing RNAs.

Key words: RNA, Thio-phosphate, Sulfur–mercury interaction, APM, Gel electrophoresis

1. Introduction

RNA can acquire sulfur by several means, both within cells and through manipulation in vitro. The modified bases of tRNA constitute the major source of natural sulfur-containing nucleotides in nature, such as thio-substitution of the keto oxygen on positions 2 or 4 of uridine bases and other derivatives (1). 4-thio-U is photoreactive, and is used in RNA–protein cross-linking studies, such as Photoactivatable-Ribonucleoside-Enhanced Crosslinking and Immunoprecipitation (PAR-CLIP) (2). Substituting one of the nonbridging phosphate oxygens with sulfur produces a phosphorothioate. Importantly, for the method presented here, the sulfur atom within phosphorothioates carries a stable negative charge, which greatly increases its chemical reactivity. Phosphorothioate substitutions in catalytic RNAs, such as the hammerhead ribozyme, have been used to analyze the catalytic activity and folding stability of these ribozymes (3, 4), and to determine whether metal ions are coordinating to specific internucleotide linkages (5, 6). Oligonucleotides with

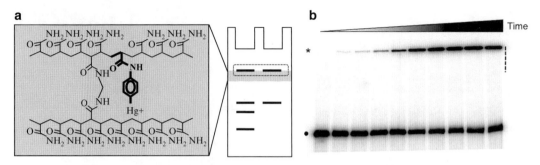

Fig. 1. Three-layered APM gel. (**a**) *Left,* schematic of a polyacrylamide matrix with one APM unit (*in bold*). *Right,* schematic of RNA migration through a three-layer gel. Thio-RNA accumulates at the interface between the top layer and the APM-containing layer (*shaded*), from which it can be excised (*dotted box*) as described below. (**b**) An example of using a three-layer APM gel to analyze accumulation of self-thio-phosphorylated RNA product (*asterisk*) by kinase ribozyme K28(1-98) (23) over time. Non-reacted RNA passes through the APM layer (*dot*). *Dashed vertical line on the right* indicates the height and position of the APM layer (100 μg/mL).

multiple internal phosphorothioate substitutions are under investigation as potential therapeutic agents, in part because of their reduced sensitivity to nucleases (7), and internal phosphorothioate substitutions are critical for nucleotide analog interference mapping (NAIM) to determine the roles of specific chemical moieties in RNA structure and function (8). 5′-thio-phosphorylated RNA is often prepared with the purpose of appending a fluorophore or some other useful compounds at the end of the molecule by virtue of the high reactivity of the sulfur atom, which serves to make maleimide or iodoacetamide conjugates (9). Thiophosphate substitutions have also been widely exploited to reveal RNA–protein contacts, such as the tRNA–protein interactions (10–13). Thus, there are many contexts in which sulfur-containing RNAs can be encountered or generated and for which it can be desirable to have a ready method of analyzing and purifying such RNAs.

Mercury was first used in polyacrylamide gels by Igloi (14) to analyze the content of thio-substituted phosphates in RNAs of different lengths and origins. Mercury is a "soft" metal ion that forms a coordinate covalent bond with the "soft" sulfur ligand, reducing the migration of RNAs that contain sulfur. The mobilities of sulfur-containing RNAs can be highly sensitive to a number of factors. As a result, this method has proven to be very useful for separating RNA molecules on the basis of the numbers of sulfurs they contain or for detecting different positions of the thiols within the RNA molecule (due to position-specific effects in mobility). However, this same sensitivity is problematic in applications, such as purifying all sulfur-containing molecules from non-modified RNAs irrespective of the number and position of the modifications. To this end, we developed methods involving three-layered polyacrylamide gels in which only the middle layer contains a high amount of the organomercurial compound (15) (Fig. 1a). All thio-containing molecules are retained at the interface between the top layer (normal polyacrylamide) and the middle, organomercurial layer, while RNAs that lack sulfur pass into the bottom layer and are separated on the basis of size. This strategy allows the facile separation of S- and non-S-substituted RNAs. In our lab, we have used this method to separate and quantify the products of in vitro-selected kinase ribozymes' self-thio-phosphorylation reactions (15–24) (Fig. 1b).

2. Materials

Prepare each solution and compound using clean, sterile, RNAse-free labware and Milli-Q water with a resistivity of 18 MΩ. We have not found it necessary to use DEPC-treated water in our RNA manipulations.

2.1. Synthesis of [(N-Acryloylamino)Phenyl]Mercuric Chloride Stock Solution (APM)

1. (*p*-aminophenyl)mercuric acetate.
2. Acetonitrile.
3. 1.2 M $NaHCO_3$.
4. Acryloyl chloride.
5. Dioxane.
6. Formamide.
7. Whatman filter paper.

2.2. Preparation of Three-Layered APM Gels

1. 1–2 mg/mL (0.25–0.50 mM) APM stock solution in formamide.
2. Acrylamide:bisacrylamide (19:1) 40% solution.
3. Urea.
4. 10× TBE (Tris–Borate–EDTA): To a 1-L beaker, add 108 g Tris base, 55 g boric acid, 40 mL 0.5 M EDTA, pH 8.0. Sterilize by autoclaving.
5. 10% Ammonium persulfate (APS) solution: Dissolve 1 g APS in 10 mL Milli-Q water, and sterilize by filtration. Store at 4°C. Do not store for more than 1–2 weeks.
6. *N*, *N*, *N'*, *N'*-Tetramethylethylenediamine (TEMED).

2.3. Phosphorimaging and Gel Analysis

No specialized materials required.

2.4. Gel Extraction of Thio-Mercury-Linked RNAs from APM-Polyacrylamide

1. 5 M Ammonium acetate, pH 5.5.
2. 0.5 M EDTA, pH 8.0.
3. 1 M DTT.
4. Pierce Centrifuge Columns, 0.8 mL.

3. Methods

Perform each procedure using clean, sterile, RNAse-free labware and solutions.

3.1. Synthesis of [(N-Acryloylamino)Phenyl]Mercuric Chloride

For the synthesis of the APM powder, we strictly follow the procedure as in ref. 11. In particular:

1. To 0.35 g of (*p*-aminophenyl)mercuric acetate, add 8 mL acetonitrile at 0°C, and stir.
2. Add 2 mL 1.2 M $NaHCO_3$. The solution will start separating into two phases.
3. While vigorously stirring, add 10-μl aliquots of acryloyl chloride up to a total of 0.2 mL added material over a period of 10 min. A white precipitate will form. Let the reaction continue overnight while stirring at 4°C.
4. Remove the white upper phase, which contains the product, and centrifuge.
5. Wash the resulting pellet with water by resuspending the pellet in water and centrifuging again.
6. Dissolve the pellet in 8.5 mL dioxane by warming up to 50°C, and rapidly filter the solution through Whatman filter paper while it is still warm (see Note 1).
7. Leave the solution at room temperature until solid crystals form, then wash the crystals with water, and dry under vacuum.
8. Transfer powder to a pre-weighed empty tube and weigh again to determine yield.
9. Store the APM powder at 4°C.
10. To use in polyacrylamide gels, dissolve in formamide to obtain a stock solution of 1–2 mg/mL. Store this solution at room temperature to avoid slow crystallization of the APM, and use as needed.

3.2. Preparation of Three-Layered APM Gels

This protocol is set for the preparation of medium-sized gels (about 18×25 cm), 0.8-mm thick. For different sizes, just adjust the amounts of solutions and compounds accordingly. Likewise, here, we report the methodology for preparing 10% polyacrylamide and 8 M urea denaturing gel solutions. For different polymer concentrations and/or denaturing strength, adjust accordingly.

1. Prepare 10%, 8 M urea polyacrylamide ready-to-use solution: In a beaker, mix 250 mL 40% acrylamide:bisacrylamide (19:1) solution, 100 mL 10× TBE, and 480 g urea. Stir vigorously, and slowly bring to volume with water while the urea is

dissolving. When the solution reaches a clear color and room temperature, sterilize by filtration. This solution will be stable at room temperature for several months (see Note 2).

2. Assemble the polyacrylamide electrophoresis gel-casting system as for standard gels, making sure to perfectly seal the bottom of the gel cast, which will need to be in an upright position at all times.

3. In a beaker, pour 40 mL of ready-to-use 10% polyacrylamide solution. Add 400 µl 10% APM and 40 µl TEMED, and mix thoroughly (see Note 3).

4. Pour the gel solution into the glass sandwich with the aid of a 50-mL pipettor or straight from the beaker, avoiding trapping of air bubbles in the gel. Fill the gel cast up to about ¾ of the total height of the gel.

5. Let the first layer polymerize in an upright, flat position so as to obtain a flat even layer. Do not allow the interface to dry out (see Note 4).

6. In a small beaker, pour 4 mL of ready-to-use polyacrylamide solution. Add the desired amount of 1–2 mg/mL APM in formamide (for a more precise APM concentration in the layer, make sure to remove from the 4 mL of polyacrylamide solution the same volume of APM you will add) (see Note 5). Mix thoroughly. Add 40 µl 10% APS and 4 µl TEMED. Mix thoroughly. Pour all 4 mL of the second gel layer onto the first layer using a 1-mL pipette, making sure to pour the APM-gel solution from one corner of the gel cast and in an even, smooth, constant flow so as to avoid mercury contamination of other parts of the gel and gel cast.

7. Let the second layer polymerize in an upright, flat position so as to obtain a flat even layer. Do not allow the interface to dry out (see Note 6).

8. Prepare the third and last layer by mixing about 15 mL of ready-to-use polyacrylamide solution with 15 µl TEMED and 150 µl 10% APS. Pour on top of the second layer and add the comb, making sure not to trap bubbles in the wells.

9. Allow the three-layered gel to polymerize for 20–60 min.

10. Pre-run and load the gel as for standard polyacrylamide gel electrophoresis.

11. For good resolution, run the gel at least until any non-thio-RNA present in the samples passes through the middle layer so as to obtain optimal separation.

3.3. Phosphorimaging and Gel Analysis

This protocol refers to ^{32}P-labeled samples, but results can be visualized by several other standard methods for polyacrylamide gel electrophoresis imaging, such as fluorescence detection and staining with ethidium bromide (see Note 7).

1. Remove the spacers from around the gel and separate the two glass plates by pulling them apart with a spatula.
2. Transfer the gel to plastic wrap (in case the thio-RNA needs to be recovered) or more easily to a used X-ray film (when imaging is all that is needed), paying attention not to rip apart the layers. In case of film, make sure to cover the other side of the gel with plastic wrap.
3. Expose the gel to (unused) film and/or to a phosphorimager screen.
4. The obtained image will look somewhat like the gel in Fig. 1b. The thio-RNAs to be analyzed will appear as bands running exactly on the upper interface of the APM layer, while any non-thiolated RNA present in the sample will run at its expected size in a normal denaturing gel.
5. If desired, ratios of thiolated versus non-thiolated RNAs in the sample can be calculated by quantifying the relative intensities of the bands within a lane, after making sure that the image signal is non-saturated. This can be achieved with any gel-imaging software.
6. If the thio-RNAs need to be recovered from the gel, proceed to Subheading 3.4.

3.4. Gel Extraction of Thio-Mercury-Linked RNAs from APM-Polyacrylamide

1. In a 15-mL conical tube, prepare 10 mL APM-gel extraction buffer: Mix 1 mL 5 M ammonium acetate, pH 5.5 (0.5 M final), 5 mL 1 M DTT (0.5 M final), 0.2 mL 0.5 M EDTA, pH 8.0 (10 mM final), and 3.8 mL water.
2. Cut the band of interest from the gel with a clean blade (see Note 8).
3. Transfer the gel slice to a clean 1.5-mL centrifuge tube, and add 500 µl APM elution buffer. Freeze and thaw the gel slice a couple of times. This partially disrupts the gel matrix and aids elution of the RNA from the gel matrix.
4. Thoroughly crush the gel slice with the silicon plunger's top of a sterile 1-mL syringe, until obtaining a smooth slurry. Note that contrary to normal RNA polyacrylamide gel extractions, in the case of APM-gels, thio-RNA elution cannot be accomplished by overnight incubations at 4°C or other passive diffusion methods due to the strong interaction between sulfur and mercury. Instead, active competition of DTT for the interaction with Hg(II) needs to be achieved by thoroughly crushing the gel slice in the presence of high concentrations of a reductant that contains mercury, such as DTT or β-mercapto-ethanol (see Notes 9–11).

5. Transfer the gel slurry to a filtered column. We use Pierce 0.8-mL centrifuge columns, which carry a polyethylene filter with 30-μm pore size.

6. Place the column in a new 1.5-mL tube, and spin at maximum speed for 2–3 min.

7. Wash the gel slurry left on the column with 200 μl of new APM elution buffer, and then repeat centrifugation.

8. Proceed to ethanol precipitation to concentrate the RNA and eliminate excess salts.

4. Notes

1. In preparing APM, when filtering the pellet dissolved in dioxane, do not use nitrocellulose filters because the organic solvent will dissolve the matrix.

2. Do not store 8 M urea polyacrylamide ready-to-use solutions at 4°C because the low temperature will cause the urea to crystallize in the refrigerator. These solutions are stable for several months at room temperature. To store polyacrylamide denaturing gel solutions at 4°C, limit the concentration of the urea in the solution to 7 M.

3. Although 10% APM solution for polyacrylamide gel polymerization should be prepared fresh as needed, we find that with the ratios we use in the lab (1:1,000) a few 10-mL aliquots can be stored at 4°C for up to a month without compromising good electrophoresis.

4. To avoid letting the gel layer interfaces dry out while casting the gel, it is possible to pour Milli-Q water *very gently* into the gel cast right after pouring the gel layer. Even if it will appear to mix at first, the water phase will soon separate and float on the gel phase, preventing it from drying out and ensuring a flat, even line to form. Once the polyacrylamide layer is polymerized, the water can be easily poured back out of the gel cast.

5. The concentration of APM to be used is different in each experiment. The APM layer can easily be saturated by overloading the wells with thio-RNA or by the presence of other free sulfur-containing molecules (for example, see Fig. 1 of ref. 15). Saturation of the layer will cause your thio-RNAs to fully or partially run through the APM layer instead of being retained at its upper interface. Remember that the mercury–sulfur interaction in a three-layered gel is a 1:1 interaction in a solid phase, confined to a specific space and number of molecules in your layer. The binding capacity is mainly determined by the APM concentration, and by the cross-sectional area of the well

(that is, the thickness of your gel and the width of your wells). In contrast, the height of the middle layer is generally not an issue. To determine which concentration best suits your needs, perform pilot experiments. A good APM concentration range for pilot experiments is 1–100 μg/mL.

6. After pouring the APM-containing layer, droplets of mercury-containing polyacrylamide could get trapped in the still-empty portion of the gel cast, causing a chaotic and spotted retardation of thio-RNAs, and sometimes difficulty in interpreting the results. To avoid this, water can be poured into the gel cast right after the APM-polyacrylamide layer, and then discarded after the layer polymerizes as described above (Note 4). Alternatively, droplets that cling to the glass can be removed through repeated rinses after the APM layer polymerizes. If stubborn droplets of polymerized APM-gel persist, a small piece of used autoradiography film can be used to remove them mechanically.

7. Imaging of three-layered APM gels can be accomplished with virtually any standard method for polyacrylamide gels. Even though radio-imaging ensures higher resolution and sensitivity, the generally low concentration of APM in the gel solution allows for general procedures, such as staining with ethidium bromide or silver.

8. Even though APM gels can be stained or imaged in several different ways, the gel extraction of thio-RNAs retained at the interface of the APM layer can also be performed "blindly." Indeed, in most cases, all the thio-RNAs can be recovered from the gel by cutting a gel slice about 3 mm above and 3 mm below the upper APM interface of the lane in which the sample was run (see dotted box in Fig. 1).

9. Compounds that carry a negatively charged sulfur interact more strongly with mercury than sulfurs that carry partial negative charges or no charge, and the number of sulfurs also influences the strength of the interaction. Thus, an RNA with a thiophosphate (full negative charge) will be bound more tightly and require more aggressive elution than an RNA with a 4-thio-U substitution (no charge on the mercaptoketone sulfur, and only partial sampling of the tautomeric form). Sulfates and disulfides do not interact appreciably with mercury. Similarly, reductants that do not contain sulfur, such as tris-carboxyethyl phosphine (TCEP) (15), are fully compatible with the three-layer APM gel method.

10. While a higher than needed concentration of APM is best to avoid layer saturation when gels are run for the purpose of monitoring and/or quantifying a phenomenon, a lower concentration may be preferable when thio-RNAs need to be

recovered from the APM interface. Indeed, in these cases, it is best to use the minimal concentration of APM that you need to accomplish the separation, as excess mercury rebinds the sulfur atoms in the RNA during the process of DTT-gel extraction and impedes recovery.

11. DTT might cause a problem in subsequent manipulations of APM-gel-extracted RNA. A simple way to get rid of excess DTT in the final sample is to dry the pellet after centrifugation for longer than usual so as to allow for the DTT to completely evaporate. Beta-mercapto-ethanol has a higher vapor pressure than DTT and is removed especially well by this drying method, although it is less effective during APM-gel extraction.

Acknowledgments

This work was supported by National Science Foundation grant CHE-1057506 to Donald H. Burke.

References

1. Sprinzl M et al (1987) Compilation of tRNA sequences and sequences of tRNA genes. Nucleic Acids Res 15(Suppl):r53–r188
2. Hafner M et al (2010) PAR-CliP–a method to identify transcriptome-wide the binding sites of RNA binding proteins. J Vis Exp 2(41):2034
3. Chowrira B, Burke J (1992) Extensive phosphorothioate substitution yields highly active and nuclease resistant hairpin ribozymes. Nucleic Acids Res 20(11):2835–2840
4. Breaker RR et al (2003) A common speed limit for RNA-cleaving ribozymes and deoxyribozymes. RNA 9:949–957
5. Dahm SC, Uhlenbeck OC (1991) Role of different metal ions in the hammerhead RNA cleavage reaction. Biochemistry 30(39):9464–9469
6. Scott EC, Uhlenbeck OC (1999) A re-investigation of the thio effect at the hammerhead cleavage site. Nucleic Acids Res 27:479–484
7. Dean NM, McKay RA (1994) Inhibition of protein kinase C-alpha expression in mice after systemic administration of phosphorothioate antisense oligodeoxynucleotides. Proc Natl Acad Sci USA 91:11762–11766
8. Cochrane JC, Strobel SA (2004) Probing RNA structure and function by nucleotide analog interference mapping. Curr Protoc Nucleic Acid Chem, Chapter 6:Unit 6.9
9. Czworkowski J, Odom OW, Hardesty B (1991) Fluorescence study of the topology of messenger RNA bound to the 30S ribosomal subunit of Escherichia Coli. Biochemistry 30:4821–4830
10. Milligan JF, Uhlenbeck OC (1989) Determination of RNA-protein contacts using thiophosphate substitutions. Biochemistry 28(7):2849–2855
11. Schatz D, Leberman R, Eckstein F (1991) Interaction of Escherichia coli tRNA(Ser) with its cognate aminoacyl-tRNA synthetase as determined by footprinting with phosphorothioate-containing tRNA transcripts. Proc Natl Acad Sci USA 88:6132–6136
12. Rudinger J et al (1992) Determinant nucleotides of yeast tRNA(Asp) interact directly with aspartyl-tRNA synthetase. Proc Natl Acad Sci USA 89:5882–5886
13. Kreutzer R, Kern D, Giegé R, Rudinger J (1995) Footprinting of tRNA(Phe) transcripts from Thermus thermophilus HB8 with the homologous phenylalanyl-tRNA synthetase reveals a novel mode of interaction. Nucleic Acids Res 23(22):4598–4602
14. Igloi G (1988) Interaction of tRNAs and of phosphorothioate-substituted nucleic acids with an organomercurial. Probing the chemical environment of thiolated residues by affinity electrophoresis. Biochemistry 27:3842–3849

15. Rhee SS, Burke DH (2004) Tris(2-carboxyethyl)phosphine stabilization of RNA: comparison with dithiothreitol for use with nucleic acid and thiophosphoryl chemistry. Anal Biochem 325:137–143
16. Saran D et al (2005) A *trans* acting ribozyme that phosphorylates exogenous RNA. Biochemistry 44:15007–15016
17. Saran D (2006) Multiple-turnover thio-ATP hydrolase and phospho-enzyme intermediate formation activities catalyzed by an RNA enzyme. Nucleic Acids Res 34:3201–3208
18. Cho B-R, Burke DH (2006) Structural stabilization of the Kin.46 self-kinasing ribozyme through topological rearrangement. RNA 12:2118–2125
19. Saran D, Burke DH (2007) Synthesis of photo-cleavable and non-cleavable substrate-DNA and substrate-RNA conjugates for the selection of nucleic acid catalysts. Bioconjugate Chem 18:275–279
20. Cho B-R, Burke DH (2007) Conformational dynamics of self-thiophosphorylating RNA. Bull Korean Chem Soc 28:463–466
21. Biondi E et al (2010) Convergent donor and acceptor substrate utilization among kinase ribozymes. Nucleic Acids Res 38(19): 6785–6795
22. Rhee SS, Burke DH (2010) Active site assembly within a kinase ribozyme. RNA 16:2349–2359
23. Biondi E, Maxwell AWR, Burke DH (2012) A small ribozyme with dual-site kinase activity. Nucleic Acids Res: in press
24. Biondi E, Sawyer AW, Maxwell AWR, Burke DH (2012) Unusual dependence on both pH and Cu^{2+} in a dual-site kinase ribozyme. *(submitted)*

Chapter 9

RNAse Mapping and Quantitation of RNA Isoforms

Lakshminarayan K. Venkatesh, Olufemi Fasina, and David J. Pintel

Abstract

The ribonuclease protection assay (RPA) has emerged as an important methodology for the detection, mapping, and quantification of RNAs. In this assay, total or cytoplasmic RNAs are hybridized to a high-specific activity antisense radioactive RNA probe synthesized by in vitro transcription from the SP6 or T7 promoter of an appropriate linearized plasmid template by the bacteriophage SP6 or T7 polymerase, respectively. The RNA hybrids are subjected to RNAse digestion and the protected products are resolved by denaturing polyacrylamide gel electrophoresis to allow detection of specific RNA fragments by subsequent autoradiography. RPAs are highly sensitive, the probes can be specifically targeted, and, when performed in probe excess, are quantitative, making them the method of choice for many analyses of RNA processing events.

Key words: In vitro transcription, SP6 or T7 RNA polymerase, High-specific activity RNA probe, RNA hybridization, RNAse A and T1 digestion, Ribonuclease protection assays, Denaturing polyacrylamide gel electrophoresis

1. Introduction

The ribonuclease protection assay (RPA) constitutes an important technique for the rapid detection, mapping, and quantitation of specific cellular mRNAs. This technique depends on the ability of ribonucleases to degrade unhybridized single-stranded RNA from a pool of cellular mRNAs while preserving intact a radiolabeled antisense RNA probe hybridized to its cognate target mRNA.

A segment of DNA containing the gene of interest, or a portion thereof, is cloned downstream of the bacteriophage SP6 or T7 promoter in the polylinker cloning site of a plasmid vector in an orientation that leads to synthesis of antisense RNA. The plasmid is cleaved with a restriction enzyme in the polylinker cloning site and the linearized plasmid is then transcribed with the appropriate bacteriophage polymerase in the presence of α-P^{32}-rNTP. The unincorporated radiolabel is removed from the RNA probe by repeated ethanol precipitation in the presence of ammonium acetate. An excess of the antisense radiolabeled RNA probe is then hybridized to cellular mRNA. After the unhybridized RNA is digested with a mixture of RNAse A and T_1 to remove free probe, the radiolabeled RNA:RNA hybrid formed by probe segments annealed to homologous sequences in the cellular sample RNA is detected by denaturing polyacrylamide gel electrophoresis and subsequent autoradiographic exposure of the gel (1, 2). Protected bands can also be easily quantified by phosphorimager analysis, and relative molar ratios of various species can be compared by standardizing for the number of radioactive nucleotides within each protected fragment. This protocol was first reported by Zinn et al. (3) and described subsequently in procedural detail by Melton et al. (4).

RPAs, which are done in solution, are more sensitive than standard Northern Blot analysis which relies on the transfer of target RNA to membranes. To maximize sensitivity, multiple radioactive nucleotides can be used to generate the RPA antisense probe if desired. Furthermore, Northern Blotting is poorly suited to fine-mapping RNA landmarks, such as initiation and termination sites and splice junctions. Standard reverse transcription-PCR (RT-PCR) can address such analyses but is semiquantitative at best, and quantitative, real-time RT-PCR, while useful for transcript quantification, is cumbersome when applied to investigation of mRNA isoforms. RPA antisense probes can be chosen to span and so map the 5′ initiation site, to span and thus map and quantify sites of polyadenylation, and to map splice donors and acceptors and thereby both map and quantify various spliced RNAs. In addition, in cases where multiple overlapping RNAs are generated from a single transcription unit, it is often possible to design a single probe that can reveal the true relative abundances of these various RNAs (see Fig. 1).

In our laboratory, we have been using RPAs for studying alternative splicing and polyadenylation of pre-mRNAs generated by parvoviruses during infection. Viruses need to maximize the expression of their compact genomes. DNA viruses make extensive use of alternative pre-mRNA processing, including alternative splicing and alternative polyadenylation, to generate multiple mRNAs which encode multiple proteins, often from overlapping open reading frames. RPAs are particularly suited to the analysis of such variation in mRNA production. We have shown this technique to be a facile method for mapping and quantifying various alternatively spliced RNAs generated by parvoviruses (6) (see Fig. 1), for understanding connections between alternative splicing and alternative polyadenylation (5, 6), for monitoring the

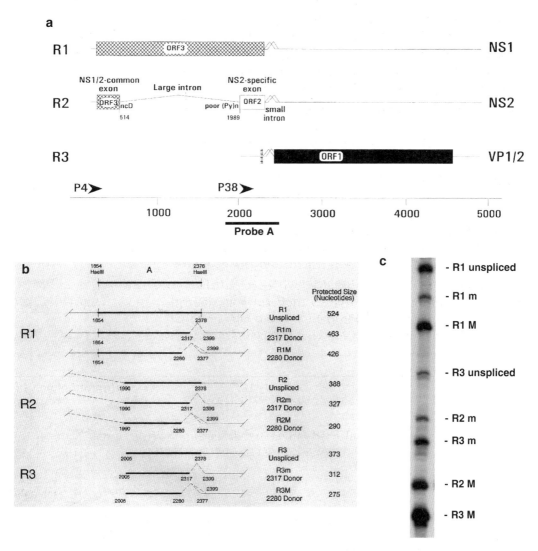

Fig. 1. A single RPA probe can quantify the relative levels of RNAs generated by the parvovirus minute virus of mice (MVM) during infection of murine cells. (**a**) Genetic map of MVM showing topography of the three major transcripts: R1 and R2 generated from a promoter at map unit 4, and R3 generated from a promoter at map unit 38. The translated open reading frames that encode NS1 and NS2 and the capsid proteins VP1 and VP2 from these mRNAs are shown. All RNAs undergo alternative splicing at a small intron in the center of the genome and are polyadenylated at the right-hand end (6, 7). The R2 mRNA is additionally spliced within the nonstructural gene region. Also shown is an RPA probe (probe A), which spans MVM nucleotides 1854–2378, and which was designed to differentially quantify the major RNA species generated. (**b**) A diagram showing the RPA probe A, spanning from nt 1854 before the large intron acceptor to nt 2378 in the middle of the small intron, and the species predicted to be protected by MVM RNAs generated during infection. (**c**) An RPA using molar excess of probe A applied to 10 μg of MVM RNA showing the various species as predicted in (**b**).

export of spliced and unspliced RNAs (5, 6), and for following splicing of viral RNAs in response to viral transactivating proteins that influence RNA processing (6).

2. Materials

Unless otherwise stated, all reagents for RNA manipulation should be treated with diethyl pyrocarbonate (DEPC), a potent RNAse decontaminant, or prepared in DEPC-treated double-distilled water (ddH$_2$O). Gloves should always be worn while handling plasticware and RNA solutions. See Notes 1–4.

2.1. Isolation of Total RNA from Mammalian Cells

1. TRIzol reagent (Invitrogen). A monophasic mixture of phenol and guanidine isothiocyanate.
2. Isopropanol.
3. 3.0 M Sodium acetate (NaOAc), pH 5.2. Adjust pH with glacial acetic acid.

2.2. Isolation of Cytoplasmic RNA from Mammalian Cells

1. DEPC-treated water: Add 1 ml of DEPC to 1 L of ddH$_2$O. Shake vigorously, stir overnight at room temperature, and autoclave.
2. Phosphate-buffered saline (PBS): 137 mM NaCl, 2.7 mM KCl, 4.3 mM Na$_2$HPO$_4$·7H$_2$O, 1.4 mM KH$_2$PO$_4$.
3. 10% Sodium dodecyl sulfate (SDS).
4. Phenol: Saturate with DEPC-treated ddH$_2$O and then equilibrate with 0.1 M Tris–Cl, pH 7.0.
5. Chloroform/isoamyl alcohol: Mix in 24:1 ratio.
6. 1.0 M Tris–HCl, pH 7.5 and 8.0: Dissolve 12.11 g Tris in 50 ml DEPC-H$_2$O, adjust pH to 7.5 and 8.0 with HCl, bring volume to 100 ml with DEPC-H$_2$O, and autoclave.
7. 0.25 M EDTA, pH 8.0.
8. Cell lysis buffer: 50 mM Tris–Cl, pH 8.0, 100 mM NaCl, 5 mM MgCl$_2$, 0.5% Nonidet P-40.
9. 5.0 M NaCl.
10. 1 M Magnesium chloride (MgCl$_2$).

2.3. In Vitro Transcription

1. 5× SP6 Transcription buffer: 200 mM Tris, pH 7.5, 30 mM MgCl$_2$, 10 mM spermidine, 50 mM NaCl. Freeze at –20°C.
2. 1 M Dithiothreitol (DTT).
3. 5 mM rNTPs: 5 mM rATP, 5 mM rCTP, 5 mM rGTP. Made from 100 mM stocks of rNTPs (Invitrogen).
4. 100 μM rUTP (Invitrogen).
5. α-P^{32}-rUTP (10 μCi/μl, New England Nuclear).
6. RNasin. Ribonuclease Inhibitor (Promega Corporation).
7. SP6 RNA Polymerase (Promega Corporation).

8. 200 mM Vanadyl ribonucleoside complex. An RNase Inhibitor. (Bethesda Research Laboratories).
9. SDS–Tris–EDTA: 0.2% SDS, 2 mM Tris, pH 8.0, 1 mM EDTA.
10. 5 mg/ml tRNA.
11. 2 M Ammonium acetate (NH_4OAc).
12. RQ1 DNAse (Promega Corporation).
13. 5× RNAse protection hybridization buffer: 200 mM Pipes, pH 6.4, 2.0 M NaCl, 5 mM EDTA.

2.4. RNase Protection Assay

1. Formamide.
2. RNAse digestion buffer: 300 mM NaCl, 10 mM Tris, pH 7.5, 5 mM EDTA.
3. RNAse A: 40 mg/ml in 10 mM Tris, pH 7.5, 10 mM NaCl. Boil for 5 min and store at −20°C.
4. RNAse T1: 2 mg/ml in 10 mM Tris–Cl, pH 7.5, 10 mM NaCl. Boil for 5 min and store at −20°C.
5. Sequencing dye: 90% formamide, 0.1% xylene cyanol, 0.1% bromophenol blue (BPB), 5 mM EDTA.

3. Methods

For some applications, isolation of total RNA is adequate; however, for analyses of mRNA export, abundance, stability, and translatability, it is often preferable to isolate cytoplasmic RNA.

3.1. Isolation of Total RNA from Mammalian Cells

1. Lyse cell monolayer in a 60-mm culture dish (approximately $1-2 \times 10^6$ cells) by adding 1 ml of Trizol reagent directly to the cells and passing the solution repeatedly through a pipette.
2. Incubate the lysate at room temperature for 5 min to allow complete dissociation of nucleoprotein complexes. Transfer the lysate to a 1.5-ml microcentrifuge tube.
3. Add 0.2 ml of chloroform and shake the tube vigorously for 15 s; incubate at room temperature for 3–5 min.
4. Centrifuge sample at $12,000 \times g$ for 15 min at 4°C. The mixture will separate into a lower red phenol–chloroform phase, an interphase, and a colorless upper aqueous phase containing the RNA.
5. Transfer the aqueous phase carefully to a new tube, and add 0.5 ml of isopropyl alcohol to precipitate the RNA. Incubate the sample at room temperature for 10 min and centrifuge at $12,000 \times g$ for 10 min at 4°C.

6. Remove the supernatant, wash the pellet once with 1 ml of 75% ethanol, and centrifuge at 12,000 × g for 5 min at 4°C.

7. Air dry the RNA pellet. Complete drying of the pellet will tend to reduce subsequent solubility of the pellet. Dissolve RNA pellet in 100 µl of RNAse-free H_2O, and incubate at 55–60°C for 10 min to completely dissolve the pellet. Take the OD_{260} of the RNA solution. Store the RNA solution at –20 to –70°C in 10-µg aliquots, adding NaOAc to 0.3 M and two volumes of ethanol (see Note 5).

3.2. Isolation of Cytoplasmic RNA from Mammalian Cells

1. Wash cell monolayer in tissue culture dish with 5 ml of ice-cold PBS. Scrape the cells into 1.5 ml PBS in a microcentrifuge tube and collect by centrifuging at 4°C for 5 min at 1500 × g.

2. Resuspend cell pellet in 375 µl ice-cold cell lysis buffer. Incubate on ice for 5 min. The cell suspension should clear due to lysis of cells.

3. Transfer the clear cell lysate to a new microcentrifuge tube. Spin for 5 min at 4°C.

4. Carefully avoiding the nuclear pellet, remove supernatant to a clean centrifuge tube containing 8 µl of 10% SDS. Vortex to mix.

5. Add 1 µl of 50 mg/ml Proteinase K. Incubate at 37°C for 15 min.

6. Extract with 400 µl phenol/chloroform/isoamyl alcohol. Vortex thoroughly for 1 min. Microcentrifuge for 5 min at room temperature to recover the upper aqueous phase.

7. Carefully transfer the upper aqueous phase containing the cytoplasmic RNA, avoiding the white interface, to a fresh microcentrifuge tube (see Note 6).

8. Extract the supernatant with 400 µl chloroform/isoamyl alcohol by vortexing for 30 s, and recover the aqueous phase by centrifuging as above.

9. Add 40 µl of 3 M NaOAc, pH 5.2, and 1 ml of ethanol. Incubate at –20°C for 1 h.

10. Collect RNA by centrifugation at 12,000 × g for 15 min at 4°C. Rinse pellet with 1 ml of 70% ethanol.

11. Dry pellet and redissolve in 100 µl H_2O. Take OD_{260} and store RNA in aliquots at –70°C by precipitating with NaOAc and ethanol as described in Subheading 3.1, step 7, above.

3.3. In Vitro Transcription

1. Combine the following reagents in the order shown after all have been brought to room temperature (see Note 7).

   Assemble the reaction in a 1.5-ml microcentrifuge tube (total reaction volume = 20 µl).

DEPC-H$_2$O	1.0 µl
5× SP6 Transcription buffer	4.0 µl
100 mM DTT	2.0 µl
5 mM rNTPs	2.0 µl
100 µM rUTP	4.0 µl
α-P^{32}-rUTP	4.0 µl
Linearized plasmid template (1 µg)	1.0 µl
RNAsin	1.0 µl
SP6 RNA polymerase	1.0 µl

Incubate the reaction mixture at 37°C for 45 min to an hour.

2. Add 1.5 µl Vanadyl ribonucleoside complex (200 mM stock) and 1 µl RQ1 DNAse to the reaction. Incubate at 37°C for 30 min.

3. Add 50 µl of SDS–Tris–EDTA solution; extract with 100 µl of phenol/chloroform/isoamyl alcohol once by vortexing for 1 min. Spin in a microcentrifuge at 12,000 ×*g* at room temperature for 5 min. Remove aqueous supernatant to a fresh tube.

4. Add 6 µl of 5 mg/ml tRNA and 75 µl of 2.0 M NH$_4$OAc to the aqueous phase, vortex, and then add 500 µl ethanol.

5. Spin at 12,000 ×*g* for 15 min at 4°C. Resuspend pellet in 200 µl of SDS–Tris–EDTA; add 20 µl of 2.0 M NH$_4$OAc and ethanol-precipitate RNA as in step 4.

6. Spin the sample in a microcentrifuge as in step 5. Resuspend the pellet in 60 µl of RNAse protection hybridization buffer.

3.4. RNAse Protection Assay

1. Spin down 10 µg of total or cytoplasmic RNA; wash the pellet in 70% ethanol. Air dry the pellet and resuspend it in 25 µl of 1× RNAse protection hybridization buffer (see Note 8).

2. Add 5 µl of SP6 or T7 polymerase-derived radiolabeled RNA probe to 25 µl of the above-described mRNA (see Note 9). Heat to 90°C for 15 min to denature the RNA; transfer to 45–51°C water bath and incubate overnight (12–16 h).

3. Add 300 µl of a 1:2,500 dilution of RNAse A and T1 in RNAse digestion buffer; incubate at 37°C for at least 45 min.

4. Add 20 µl of 10% SDS and 1 µl of 50 mg/ml Proteinase K. Incubate at 37°C for 15 min or at room temperature for 30 min.

5. Extract once with phenol/chloroform/isoamyl alcohol, and extract once with chloroform/isoamyl alcohol. Spin in a micro-

centrifuge at room temperature after each of these extractions to recover upper aqueous phase.

6. Add 8 μl of 5 mg/ml tRNA to the aqueous phase and precipitate with 2 volumes of ethanol.

7. Spin the preparation at $12,000 \times g$ for 15 min at 4°C. Wash the pellet with 75% ethanol and dry the pellet in a Speed-Vac for 5–10 min. Resuspend the pellet in 13 μl of RNA-sequencing dye.

8. Heat the sample at 95°C for 5 min, cool to room temperature, and load 6 μl on an 8% urea–polyacrylamide denaturing sequencing gel; run the gel at 60 W constant power (about 1,500 V) until the BPB dye reaches the bottom of the gel. The BPB comigrates in position with a 20-nucleotide fragment. Expose gel by autoradiography or image scan the gel using the Typhoon FLA 9000 phosphorimager protocol.

4. Notes

1. Unless otherwise stated, all solutions for RNA manipulation should be treated with 0.1% DEPC and autoclaved or prepared with double-distilled water that has been treated with DEPC and autoclaved. DEPC is a potent RNAse decontaminant; add 1/1,000 volume of DEPC to the solution, shake vigorously, and add a flamed stir bar. Stir the solution at room temperature overnight. Autoclave the DEPC-treated solution for 30 min to remove traces of DEPC. DEPC should be handled with care as it is a suspected carcinogen and a skin/respiratory system irritant.

2. Since bare hands often contain bacteria and mold and can thus be a source of RNAse contamination, always wear gloves while handling RNA or solutions used for RNA manipulation.

3. Reserve autoclaved plasticware (including blue or yellow pipette tips, microcentrifuge tubes, etc. that have been loaded into containers wearing gloves) for exclusive use with RNA.

4. Always clean the barrels of the pipettors with 95% ethanol and dry with a Kimwipe before using them for RNA work. Wash in 95% ethanol and flame any stir bars, spatulas, and forceps used to make RNA solutions or in any RNA protocols.

5. Always store RNA preparations in EtOH (1/10 volume 3 M NaOAc, pH 5.2, plus 2 volumes EtOH) at −20 or −70°C for long-term stability.

6. After proteinase K treatment, only a small amount of precipitate is visible at the interface. If a large amount of precipitate is

present and recovery of the aqueous phase is difficult, discard the organic phase at the bottom of the tube and add 400 μl of chloroform/isoamyl alcohol. Vortex vigorously and spin for 5 min at room temperature. The precipitate should largely disappear and the aqueous phase should be easily recoverable.

7. Warm the ingredients for the in vitro transcription protocol to room temperature prior to setting up the reaction; otherwise, the spermidine in the 5× transcription buffer may precipitate the DNA in the cold.

8. The amount of unlabeled cellular mRNA required for successful detection of an mRNA species depends on the concentration of the mRNA species and the specific activity of the RNA probe. For RNA probes labeled to high-specific activity ($>10^9$ cpm/μg), 10 μg of total RNA is sufficient to detect mRNA species present at 1–5 copies/cell (ref. 2).

9. $1-5 \times 10^5$ cpm of radiolabeled probe is sufficient for each hybridization reaction. Using our protocol for in vitro transcription, the amount of probe generated per transcription reaction is sufficient for 12–24 hybridization reactions.

Acknowledgments

Work in our laboratory has been funded by multiple grants from NIAID, NIH, to DJP.

References

1. Gilman M (1997) Ribonuclease protection assay. In: Ausubel FM, Brent R, Kingston RE, Moore DD, Seidman JG, Smith JA, Struhl K (eds) Current protocols in molecular biology. Wiley Interscience, New York, NY, p 4.7.1
2. Sambrook J, Fritsch EF, Maniatis T (1989) Mapping of RNA with ribonuclease and radiolabeled RNA probes. In: Sambrook J, Fritsch EF, Maniatis T (eds) Molecular cloning: a laboratory manual, vol 2. Cold Spring Harbor Laboratory, Cold Spring Harbor, NY, p 7.71
3. Zinn K, DiMaio D, Maniatis T (1983) Identification of two distinct regulatory regions adjacent to the human β-interferon gene. Cell 34:865–879
4. Melton DA, Krieg PA, Rebagliati MR, Maniatis T, Zinn K, Green MR (1984) Efficient in vitro synthesis of biologically active RNA and RNA hybridization probes from plasmids containing a bacteriophage SP6 promoter. Nucl Acids Res 12:7035–7056
5. Qiu J, Pintel D (2008) Processing of adeno-associated virus RNA. Front Biosci 13:3101–3115
6. Qiu J, Yoto Y, Tullis G, Pintel DJ (2006) Parvovirus RNA processing strategies. In: Kerr JR, Cotmore SF, Bloom ME, Linden RM, Parrish CR (eds) Parvoviruses, Hodder Arnold, London, UK, pp 253–274
7. Zhao Q, Schoborg RV, Pintel DJ (1994) Alternative splicing of pre-mRNAs encoding the non-structural proteins of minute virus of mice is facilitated by sequences within the downstream intron. J Virol 60:2849–2859

Chapter 10

Detection and Quantification of Viral and Satellite RNAs in Plant Hosts

Sun-Jung Kwon, Jang-Kyun Seo, and A.L.N. Rao

Abstract

Northern blotting is a valuable method for detection and quantification of RNA in the field of virology. Although many methods including a various versions of polymerase chain reaction have been developed over the years, Northern blotting has been still considered as a useful and effective method for the analysis of progeny RNA accumulation for viral and subviral pathogens, such as satellite RNAs, in plant hosts. Here, we describe a detailed Northern blot protocol for efficient detection and quantification of viral and satellite RNAs from plant hosts.

Key words: Northern blot, Viral RNA, Satellite RNA, RNA detection and quantification

1. Introduction

Regulation of genome replication is critical for successful establishment of infection by single-strand, positive-sense RNA viruses in a susceptible host (1). After entry into a cell, viral genomic RNAs are translated to yield proteins that actively participate in virus replication, encapsidation, and cell-to-cell and long-distance movement. Following the assembly of a functional replicase (virus-encoded proteins and host factors), it recruits viral genomic RNAs to specialized membrane-derived structures to initiate the synthesis of complementary negative-strand RNA. At some point, negative-strand synthesis ceases and the viral replicase uses negative-strand RNAs as templates for asymmetric synthesis of new genomic positive-sense RNA. These newly synthesized viral genomic RNAs are encapsidated by the virus structural proteins to form stable infectious virions. Each of these events are likely to involve complex macromolecular interactions because switching from (1) translation to transcription, (2) negative-strand synthesis to positive-strand synthesis, and (3) translation to encapsidation are temporally regulated during the viral life cycle (2–5).

An essential prerequisite in understanding the viral biology is to sensitively detect and accurately quantitate the accumulation levels of viral RNA at each step of virus life cycle. Advent of recombinant DNA has allowed the development of various techniques for RNA detection and quantification. Some of these include Northern blot hybridization, reverse transcription-polymerase chain reaction (RT-PCR), semi-quantitative PCR, and real-time RT-PCR (6–9). However, inherent disadvantages in optimizing conditions for specificity and "false priming" (10, 11) make the RT-PCR-based techniques unreliable for quantification of viral RNAs.

Northern blotting is a classical molecular biology method for analyzing RNA expression levels in cells or tissues (12). It reveals information regarding the size and relative abundance of RNA in a given viral infection. Despite advances made in developing RT-PCR-based sensitive methods of detection, Northern blotting is still a popular method of choice for the detection and analysis of RNA accumulation during viral replication. Especially, some RNA viruses replicate and express their genes via subgenomic RNA synthesis [e.g., Brome mosaic virus (BMV)] or generate defective interfering (DI) RNA during their life cycle (13–15). Furthermore, some RNA viruses like BMV have segmented RNA genomes (15, 16). The relative abundance and size of each genome segment or DI RNA can reliably be analyzed by Northern blotting. Here, we describe a detailed Northern blot protocol for the detection and quantification of viral RNA accumulation in infected plant cells using BMV, having a multipartite RNA genome, and a *satellite RNA of Cucumber mosaic virus* (CMV) as experimental systems.

2. Materials

2.1. RNA Extraction

1. TRIZOL® reagent (Invitrogen).
2. Chloroform.
3. Isopropyl alcohol.
4. 75% Ethanol (in nuclease-free water).
5. Nuclease-free water.

2.2. In Vitro RNA Transcription

1. Restriction enzyme.
2. Phenol:chloroform:isoamyl alchohol (25:24:1).
3. 100% ethanol.

4. 3 M Sodium acetate, pH 5.2.
5. 75% Ethanol (in nuclease-free water).
6. T7 RNA polymerase (Ambion).
7. Ribonucleotide (rNTP) mixture (Promega).
8. Ribonuclease inhibitor (NEB).
9. Nuclease-free water.
10. 7.5 M ammonium acetate.
11. 10 mg/mL yeast tRNA.
12. TE buffer: 10 mM Tris–HCl, pH 8.0, 1 mM EDTA.

2.3. Denaturing Agarose Gel Electrophoresis

1. Agarose.
2. Formaldehyde.
3. Formamide.
4. 10× MOPS buffer: Add 41.8 g of MOPS, 6.7 mL of 3 M sodium acetate, pH 5.2, and 20 mL of 0.5 M of EDTA, pH 8.0, to 800 mL nuclease-free water. Adjust pH to 7.0 with NaOH and the final volume to 1 L. Store at room temperature.

2.4. RNA Sample Preparation

1. Denaturation buffer: 1× MOPS, 18%(v/v) formaldehyde, 50%(v/v) formamide (e.g., to prepare 1 mL, mix the following: 0.5 mL of formamide, 0.18 mL of formaldehyde, 0.1 mL of 10× MOPS buffer, 0.22 mL of water).
2. 10× Gel-loading dye: 50% (v/v) glycerol, 10 mM EDTA, pH 8.0, 0.25% (w/v) bromophenol blue, 0.25% (w/v) xylene cynol FF.

2.5. Vacuum Blotting

1. VacuGene™ XL Vacuum blotting System (GE Healthcare).
2. Magnacharge Nylon membrane (GE Water & Process Technologies).
3. Whatman 3 M filter paper.
4. 20× SSC: Add 175.3 g of NaCl and 88.2 g of sodium citrate to 800 mL nuclease-free water. Adjust pH to 7.0 with HCl and adjust final volume to 1 L. Store at room temperature.
5. Ultraviolet (UV) cross-linker (Fisher Scientific).
6. Methylene blue staining solution: 0.04% (w/v) of methylene blue, 0.5 M sodium acetate, pH 5.2.

2.6. Radio-Labeled RNA Probe (Riboprobe)

1. A transcriptional plasmid with desired cDNA insert.
2. 5× Transcription buffer; 100 mM Tris–HCl, pH 8.3, 250 mM KCl, 25 mM $MgCl_2$, BSA (1 mg/mL).
3. 0.1 M DTT.

4. RNA guard.
5. T7 or T3 RNA polymerase.
6. 2.5 mM rA, G, C mixture.
7. 0.1 µM rUTP.
8. [^{32}P]UTP.

2.7. Hybridization

1. 50× Denhardt's soulution: Add 10 g of Ficoll-400, 10 g of bovine serum albumin, and 10 g of polyvinylpyrrolidone to 1 L nuclease-free water. Filter the solution through a 0.2-µM filter. Store at −20°C.
2. 10 mg/mL denatured salmon sperm DNA.
3. 50 mg/mL yeast RNA.
4. 10% SDS.
5. 5 M NaCl.
6. Nuclease-free water.
7. Prehybridization buffer: Deionized formamide 10.0 mL, SDS (10%) 2.0 mL, Denhardt's solution (50×) 2.0 mL, denatured salmon sperm DNA (10 mg/mL) 0.3 mL, NaCl (5.0 M) 4.0 mL, sterile distilled water 1.7 mL.
8. Hybridization buffer: Prehybridization buffer containing 200 µg/mL yeast RNA and appropriate amount of the ^{32}P-labeled riboprobe.
9. Hybridization oven.
10. Wash solution.
11. Phosphoimager screen.

3. Methods

3.1. RNA Extraction

1. Preparation of high-quality total RNA is a prerequisite for reliable detection and quantification of viral or satellite RNAs in the host tissue. Several commercial reagents for purifying RNA preparation, such as TRIZOL (Invitrogen) or TRI Reagent (Sigma), are available. The RNA extraction should be followed according to manufacturer's instructions. Here, we describe the use of TRIZOL reagent for RNA extraction (see Note 1).
2. Homogenize the plant tissue (100 mg) thoroughly in liquid nitrogen, add 1 mL of TRIZOL reagent, then pipette repeatedly, and transfer to a microfuge tube (see Note 2).
3. Vortex tubes containing the homogenate for 15 s and incubate at room temperature for 3 min.
4. Centrifuge the homogenate at $12,000 \times g$ for 10 min at 4°C.

5. Transfer the supernatant to a fresh microfuge tube; add 200 μL of chloroform per 1 mL TRIZOL reagent and vortex for 15 s. Incubate at room temperature for 3 min.

6. Centrifuge at 12,000×g for 15 min at 4°C.

7. Transfer the aqueous phase into a fresh microfuge tube.

8. Precipitate RNA by adding 500 μL of isopropyl alcohol, vortex briefly, then incubate at room temperature for 10 min, and centrifuge at 12,000×g for 10 min at 4°C.

9. Remove the supernatant and wash the pellet with 1 mL 75% ethanol. Centrifuge at 7,500×g for 5 min at 4°C. Carefully remove the ethanol solution, air dry the pellet, and resuspend in nuclease-free water.

10. Determine the RNA concentration by a spectrophotometer measuring the optical density at 260 nm (1 OD_{260} = 40 μg/mL).

3.2. RNA Transcription for Quantification of Viral or Satellite RNA

1. To accurately quantify the viral or satellite RNAs under investigation, it is imperative to co-electrophorese known RNAs as size markers along with the total RNA sample preparations,

2. To prepare marker RNAs, clone full-length viral RNA(s) or satellite RNA into a plasmid containing T7 promoter (see Note 3).

3. Linearize the plasmid with appropriate restriction enzyme to generate a blunt end or 5′ protruding end, (e.g., *Bam*HI; avoid using restriction enzymes that are known to produce 3′ overhangs, e.g., *Pst*I) followed by phenol extraction and ethanol precipitation. After a 75% ethanol washing, resuspend the pellet in nuclease-free water.

4. Set up an in vitro transcription reaction by adding 2 μL of 10× buffer (Ambion), 1 μL of T7 RNA polymerase (20 U/μL), optimal amount of linearized DNA template (~1 μg/μL), 4 μL of 2.5 mM rNTP mixture, and make up the volume to 20 μL with nuclease-free water. Incubate for 1 h at 37°C. Place the tube containing the reaction products on ice.

5. To remove unincorporated nucleotides, precipitate RNA transcripts by adding 80 μL of nuclease-free water, 1/2 vol of 7.5 M ammonium acetate, 1 μL of carrier yeast RNA (10 mg/mL), and 2.5 vol of cold ethanol.

6. Incubate at –80°C for 30 min and centrifuge at 4°C for 20 min.

7. Wash the pellet with 75% ethanol, dry in a speed-vac, and suspend in 50 μL of TE buffer.

8. Estimate the RNA concentration by a spectrophotometer measuring the optical density at 260 nm (1 OD_{260} = 40 μg/mL).

3.3. Denaturing Gel Electrophoresis of RNA Samples

1. Prepare 1.2% agarose gel (e.g., for 200 mL of agarose gel solution, use 2.4 g of agarose and 174 mL of nuclease-free water).
2. Microwave until the agarose is completely dissolved.
3. Cool the gel solution to approximately 60°C.
4. Add 20 mL of 10× MOPS and 6 mL of 37% formaldehyde, thoroughly mix, and pour into the gel tray.
5. Allow the gel to harden.
6. Place the gel tray in the tank containing running buffer (1× MOPS).

3.4. Electrophoresis of Viral or Satellite RNA

3.4.1. RNA Sample Preparation

1. In a sterile microfuge tube, add a known amount of total RNA (e.g., 1–10 μg) and adjust volume to 10 μL. Then, add 20 μL of denaturation buffer.
2. Vortex the mixture and centrifuge briefly.
3. Incubate the mixture at 65°C for 15 min.
4. Cool the mixture on ice, add 3 μL of 10× gel loading dye, and load each sample into the well.
5. As a marker for quantification, load quantified mixture of RNA transcripts (e.g., BMV RNAs 1, 2 and 3).
6. Electrophorese samples initially for 30 min at 5 V/cm.
7. Then, increase voltage to 10 V/cm and electrophorese until the bromophenol blue dye front is approximately 1 cm from the bottom of the gel.
8. To prevent the formation of pH gradient during electrophoresis, circulate the buffer with stir bars.

3.4.2. Transfer of RNA to Nylon Membrane by Vacuum Blotting

1. Cut a charged nylon membrane (GE Water & Process Technologies) to the size of the gel and pretreat by wetting the membrane with distilled water followed by soaking with 7× SSC.
2. Prepare the vacuum-blotting unit (e.g., VacuGene, GE Healthcare) following manufacturer's instructions.
3. Connect the blotting unit to a vacuum pump and apply the vacuum between 70 and 80 mbar.
4. Perform vacuum blotting for 2 h.
5. After blotting, dry the membrane at 55°C and place the membrane in a UV cross linker (Fisher scientific) to fix the RNA to the membrane.
6. To verify complete transfer of RNA to membrane and ensure equal loading of RNA samples, stain the membrane with methylene blue staining solution until RNA bands appear and then destain with distilled water (Fig. 1).

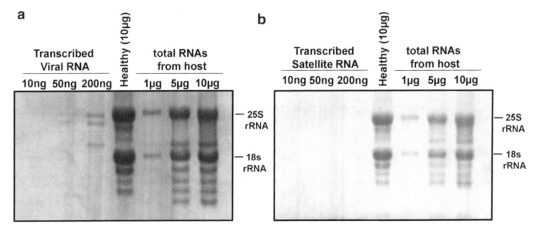

Fig.1. Images of methylene blue-stained membranes. Following RNA transfer, membranes containing either viral (**a**) or satellite RNA (**b**) were stained with methylene blue. The positions of 25- and 18-s cellular RNAs are indicated to the *right*. The amount of in vitro-synthesized RNA transcript or total RNA from leaf tissue is shown above each lane.

7. Scan the membrane to record loading controls and then dry at room temperature (see Note 4).

3.4.3. Preparation of Radiolabeled RNA Probe

1. To detect RNAs specific for either virus or satellite, clone the full-length or part of the viral/satellite genome into a plasmid containing T7 promoter to yield an antisense transcript for detecting the viral or satellite RNAs (see Note 5).

2. Linearize the DNA template with a restriction enzyme that generates 5′ overhang, as described above.

3. In a sterile microfuge tube, mix the following components:

    Transcription buffer (5×), 4.0 μL

    DTT (0.1 M), 2.0 μL

    RNA guard (40 U/mL), 1.0 μL

    ATP, CTP, GTP (2.5 mM each), 4.0 μL

    UTP (100 μM), 2.4 μL

    [^{32}P]UTP (~3,000 Ci/mmol), 3.0 μL

    DNA template linearized (1 μg/μL), 1.0 μL

    T7 or T3 RNA polymerase (25 U/μL), 1.0 μL

    Water, 1.6 μL

4. Mix the contents by vortexing and briefly centrifuge.
5. Incubate the reaction mixture at 37°C for 1 h.
6. Terminate the reaction by adding 20 μL of TE buffer.
7. Extract once with an equal volume of phenol/chloroform.

138   S.-J. Kwon et al.

8. Collect the supernatant and add 1.0 µL of carrier yeast RNA (10 mg/mL), 1/2 volume of 7.5 M ammonium acetate, and 2.5 volume of cold ethanol.

9. Incubate at −80°C for 1 h.

10. Centrifuge the tubes at 4°C for 20 min.

11. Wash the pellet with 75% ethanol, dry, and suspend in 50 µL of water.

12. Use the probes immediately or store at −80°C for later use.

3.4.4. Hybridization

1. Prehybridize the membrane in prehybridization buffer for 5–16 h at 65°C.

2. Discard the prehybridization buffer.

3. Add hybridization buffer containing the radiolabeled probe and hybridize the membrane for 16 h at 65°C.

4. Wash the membrane twice in 2× SSC and 0.1% SDS for 15 min at room temperature, and then wash twice for 30 min at 65°C in 0.2× SSC and 0.1% SDS.

5. After drying, wrap the membrane with saran wrap and expose to X-ray film or phosphorimaging screen (Fig. 2).

6. Analyze the accumulation of viral (Fig. 2a) or satellite RNA (Fig. 2b).

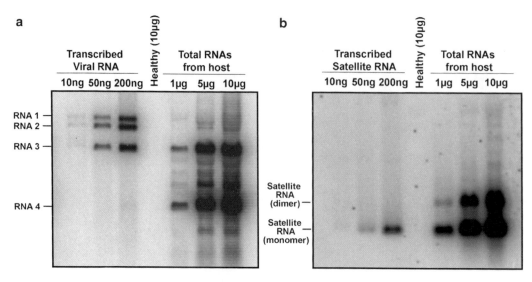

Fig. 2. Detection of viral and satellite RNA by Northern blot analysis. Blots shown in (**a**) and (**b**) are same as those shown in Fig. 1. Blot (**a**) was hybridized with a riboprobe complementary to a highly conserved 3′ tRNA-like structure sequence to specifically detect BMV RNAs. The positions of four BMV RNAs are indicated to the *left*. Blot (**b**) was hybridized with a riboprobe complementary to satellite RNA. Positions of monomeric and dimeric forms of CMV satellite RNA are indicated to the *left*.

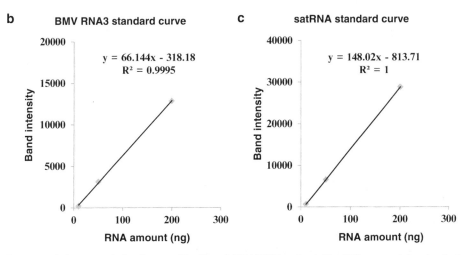

Fig. 3. A representative example for the quantification of BMV RNA3 and satellite RNA accumulation in plant cells by Northern blot-based RNA quantification assay. (**a**) Table showing raw band intensity values. Band intensity values were determined in a fluorescence plate reader using ImageQuant program. † Band intensity values of 1 μg total RNA samples were used to calculate the accumulation levels of BMV RNA3 and satellite RNA (see Note 7). (**b**, **c**) Standard curves generated using BMV RNA3 (**b**) or satellite RNA (**c**) as controls. The standard curve equations deducted from the control band intensity values were used to calculate BMV RNA3 or satellite RNA accumulation levels.

3.5. Quantification of Viral or Satellite RNA

For accurate quantification, standard curves should be generated using the quantified RNA transcripts as controls for specificity and sensitivity (see Fig. 3 for examples of absolute quantification of BMV genomic RNA3 and CMV satellite RNA monomer).

1. Calculate the intensity of each band using either ImageQuant program or any other suitable programs (Fig. 3a).

2. For using the band intensity values in RNA controls, generate a standard curve by plotting band intensity versus the amount of RNA control (Fig. 3a, b). Use a linear fit to generate a standard curve equation (see Note 6).

3. Estimate band intensity values for the samples under investigation (see Note 7) to RNA amounts using the standard curve equation determined in step 2 in Subheading 3.5.

4. Notes

1. Exercise care while extracting total RNA from plant tissue to avoid RNase contamination by wearing disposable gloves throughout experimentation.
2. Tissues must be frozen or processed immediately after harvesting from the plant.
3. Several plasmids are commercially available containing specific promoters, such as SP6, T7, and T3. Generally, each promoter region is flanked by a polylinker cloning site. If a gene of interest were fused to the 3′ end of a promoter, use of the linearized recombinant plasmid as a template in an in vitro transcription reaction by a specific RNA polymerase (e.g., T7 or T3) would result in the generation of an RNA transcript complementary to the cloned gene.
4. By placing the destained membrane between 3-MM paper, it can be stored at −20°C until further use.
5. Radiolabeled DNA probes, such as PCR product or oligonucleotide, can also be used in Northern hybridization. However, RNA–RNA hybrids are more stable than RNA–DNA hybrids and permit high stringency washes to minimize background. Unlike DNA probes, because of their single-stranded nature, RNA probes do not need to reassociate.
6. Accurate quantification of viral or satellite RNA accumulation levels in a given sample is dependent on the accuracy of the standard curve generated with controls. Thus, it is necessary to verify the coefficient of determination (R^2) of the determined standard curve equation.
7. The band intensity values do not increase proportionally when RNA samples are overloaded. Thus, using the band intensity values of the samples ranged within the standard curve is necessary for accurate determination of RNA accumulation levels.

Acknowledgments

This study was supported by a grant from the National Institutes of Health (1R21AI82301) to A.L.N. Rao.

References

1. Noueiry AO, Ahlquist P (2003) Brome mosaic virus RNA replication: revealing the role of the host in RNA virus replication. Annu Rev Phytopathol 41:77–98
2. Gamarnik AV, Andino R (1998) Switch from translation to RNA replication in a positive-stranded RNA virus. Genes Dev 12:2293–2304
3. Lemm JA, Rumenapf T, Strauss EG, Strauss JH, Rice CM (1994) Polypeptide requirements for assembly of functional Sindbis virus replication complexes: a model for the temporal regulation of minus- and plus-strand RNA synthesis. EMBO J 13:2925–2934
4. Marsh LE, Huntley CC, Pogue GP, Connell JP, Hall TC (1991) Regulation of (+):(−)-strand asymmetry in replication of brome mosaic virus RNA. Virology 182:76–83
5. Annamalai P, Rao AL (2007) In vivo packaging of brome mosaic virus RNA3, but not RNAs 1 and 2, is dependent on a cis-acting 3′ tRNA-like structure. J Virol 81:173–181
6. Wang WK, Sung TL, Tsai YC, Kao CL, Chang SM, King CC (2002) Detection of dengue virus replication in peripheral blood mononuclear cells from dengue virus type 2-infected patients by a reverse transcription-real-time PCR assay. J Clin Microbiol 40:4472–4478
7. Purcell MK, Hart SA, Kurath G, Winton JR (2006) Strand-specific, real-time RT-PCR assays for quantification of genomic and positive-sense RNAs of the fish rhabdovirus, infectious hematopoietic necrosis virus. J Virol Methods 132:18–24
8. Komurian-Pradel F, Perret M, Deiman B, Sodoyer M, Lotteau V, Paranhos-Baccala G, Andre P (2004) Strand specific quantitative real-time PCR to study replication of hepatitis C virus genome. J Virol Methods 116:103–106
9. Gu C, Zheng C, Shi L, Zhang Q, Li Y, Lu B, Xiong Y, Qu S, Shao J, Chang H (2007) Plus- and minus-stranded foot-and-mouth disease virus RNA quantified simultaneously using a novel real-time RT-PCR. Virus Genes 34:289–298
10. Lanford RE, Chavez D, Chisari FV, Sureau C (1995) Lack of detection of negative-strand hepatitis C virus RNA in peripheral blood mononuclear cells and other extrahepatic tissues by the highly strand-specific rTth reverse transcriptase PCR. J Virol 69:8079–8083
11. Lanford RE, Sureau C, Jacob JR, White R, Fuerst TR (1994) Demonstration of in vitro infection of chimpanzee hepatocytes with hepatitis C virus using strand-specific RT/PCR. Virology 202:606–614
12. Sambrook J, Russell DW (2001) Molecular cloning: a laboratory manual, 3rd edn. Cold Spring Harbor Laboratory Press, Cold Spring Harbor, NY
13. Miller WA, Koev G (2000) Synthesis of subgenomic RNAs by positive-strand RNA viruses. Virology 273:1–8
14. Simon AE, Roossinck MJ, Havelda Z (2004) Plant virus satellite and defective interfering RNAs: new paradigms for a new century. Annu Rev Phytopathol 42:415–437
15. Levy JA, Fraenkel-Conrat H, Owens RA (1994) Virology, 3rd edn. Prentice Hall, Englewood Cliffs, NJ
16. Dimmock NJ, Easton AJ, Leppard K (2007) Introduction to modern virology, 6th edn. Blackwell, Malden, MA

Chapter 11

In Situ Detection of Mature miRNAs in Plants Using LNA-Modified DNA Probes

Xiaozhen Yao, Hai Huang, and Lin Xu

Abstract

MicroRNAs (miRNAs) play important roles in development in plants, and some miRNAs show developmentally regulated organ- and tissue-specific expression patterns. Therefore, in situ detection of mature miRNAs is important for understanding the functions of both miRNAs and their targets. The construction of promoter–reporter fusions and examination of their in planta expression have been widely used and the results obtained thus far are rather informative; however, in some cases, the length of promoter that contains the entire regulatory elements is difficult to determine. In addition, traditional in situ hybridization with the antisense RNA fragment as the probe usually fails to detect miRNAs because the mature miRNAs are too short (~21 nt) to exhibit stable hybridization signals. In recent years, the locked nucleic acid (LNA)-modified DNA probe has been successfully used in animals and plants to detect small RNAs. Here, we describe a modified protocol using LNA-modified DNA probes to detect mature miRNAs in plant tissues, including the design of LNA probes and detailed steps for the in situ hybridization experiment, using *Arabidopsis* miR165 as an example.

Key words: Plant miRNA, In situ hybridization, LNA-modified DNA probe

1. Introduction

MicroRNAs (miRNAs) are approximately 21-nt noncoding RNAs that are processed from longer pri-miRNAs with an imperfect stem-loop structure (1). miRNAs were demonstrated to play fundamental roles in animal and plant development by posttranscriptionally regulating the expression of target genes (1, 2). Recent results also showed that long miRNAs could direct DNA methylation of target genes, thereby regulating the genes at the transcriptional level (3, 4). Many plant miRNAs exhibit spatial and temporal accumulation patterns during plant development, and such patterns are likely controlled by complex genetic networks. To understand miRNA functions and, more importantly, the underlying regulatory networks for miRNA accumulation, in situ analysis of miRNA distribution is an indispensable strategy.

Different methods have been used to detect miRNA distribution in recent years. Quantitative reverse transcription-polymerase chain reaction (qRT-PCR) and northern blot assays are generally used to detect expression of miRNAs in many types of tissues (5–9). RT-PCR is usually used to detect pri-miRNA, while northern blotting is used to analyze accumulation of mature miRNAs. One of the shortcomings of these two methods is that they fail to distinguish the specific cell types that express miRNAs, and the analyses usually require a relatively large amount of plant material. Increasing information concerning the transcriptional start sites of the miRNA genes has become available; thus, fusing a reporter gene to the promoter of an miRNA gene and analyzing the reporter expression in transgenic plants is a useful method for understanding miRNA distribution (10–16). However, occasionally, the reporter signals are influenced by the context of the insertion site such that the resultant expression pattern may not precisely reflect that of the endogenous gene. In addition, it is difficult to determine whether an miRNA promoter fragment used in promoter–reporter fusion contains all the necessary elements for proper expression of miRNA genes. In situ detection of RNA by hybridization with complementary probes is a widely used technique for analyzing temporal and spatial expression of RNA. However, it is not suitable for miRNA distribution because the shorter miRNA sequences are difficult to detect using an antisense RNA probe, and usually result in a low signal-to-noise ratio, especially for low-abundance miRNAs.

Locked nucleic acid (LNA) oligonucleotides are a class of bicyclic RNA analogs (17). By forming an O2′, C4′-methylene bridge, the furanose ring of the LNA sugar-phosphate backbone is chemically locked in an RNA-mimicking conformation (Fig. 1), resulting in strong thermal affinity when hybridized with complementary DNA and RNA molecules (18–21).

Probe design is critical for successful target detection. Therefore, the high thermal stability of LNAs provides ideal probes to detect small molecule RNAs. For example, using LNA-modified oligonucleotide probes, the specificity and sensitivity of miRNA detection by northern blotting were significantly improved (22). LNA-modified DNA probes have been successfully used to detect miRNA signals in zebra fish and mouse embryos (23, 24). This method was first applied in plant miRNA analysis by Valoczi et al. (25). Here, we describe a modified in situ hybridization protocol using LNA-modified DNA probe to detect miRNA accumulation in plants. The in situ hybridization steps are also modified from a previously reported protocol (26).

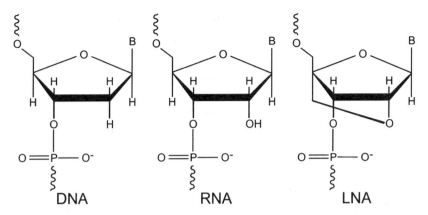

Fig. 1. The chemical structure of the LNA oligonucleotide (according to ref. 21). *B* base.

2. Materials

Preparation of all solutions for the following experiments must use diethylpyrocarbonate (DEPC)-treated water. All cylinders, beakers, and flasks need to be baked for more than 5 h at 180°C, and all centrifuge tubes and pipets should be RNase free. All containers used for incubation during hybridization or washing of specimens need to be soaked in 3% hydrogen peroxide solution overnight.

2.1. LNA-Modified DNA Probes

Both antisense and sense probes are essentially required in the in situ hybridization. It has been proved that substitution of each third nucleotide with the LNA will efficiently improve the hybridization thermal stability of probes during hybridization of corresponding RNA or DNA targets (22, 27). To follow this rule, we designed the miR165 probe as described in Fig. 2. The sequence of mature miR165 is 5′-UCGGACCAGGCUUCAUCCCCC-3′, and the designed LNA-modified DNA antisense and sense probes are as follows: 5′-GgGGgATgAAgCCtGGtCCgA-3′ for antisense and 5′-TcGGaCCaGGcTTcATcCCcC-3′ for sense probes, where the lowercase letters represent LNAs. Our LNA probes used for detection of miR165 were ordered from TaKaRa.

2.2. Solutions for Probe Labeling

1. DEPC-treated water is made by adding 1 ml of DEPC to 1,000 ml of distilled water. After being kept at 37°C overnight, the DEPC-treated water is autoclaved at 121°C for 20 min. After cooling down to the room temperature, the DEPC-treated water is autoclaved once more at 121°C for 20 min.

2. Dissolve the LNA-modified DNA probe in water at a concentration of 200 pmol/μl. This is a 10× stock solution, which is usually dispensed into aliquots and is good for at least 6 months

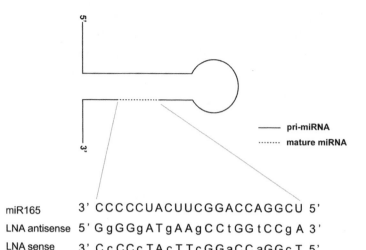

Fig. 2. Design of antisense and sense probes by using LNA oligonucleotides for in situ detection of miR165. The lowercase letters represent LNAs.

at −20°C. The 1× probe solution is made by adding 18 μl of H_2O to 2 μl of 10×-ml stock solution in a 1.5-ml Eppendorf tube, which can be used for hybridization of a total of 80 slides.

3. The digoxigenin (DIG) Oligonucleotide 3′-end labeling kit is bought from Roche company (03353575910), which contains 5× reaction buffer, 25 mM $CoCl_2$, 400 U/μl Terminal transferase, and 1 mM DIG-ddUTP.

4. 0.5 M EDTA-Na, pH 8.0.

5. 0.2 M EDTA-Na, pH 8.0.

2.3. Section Preparations

1. Several small glass bottles (about 10 ml in volume) and a vacuum air pump.

2. FAA solution: 50% ethanol, 5% acetic acid, and 3.7% formaldehyde.

3. A series concentration of ethanol: 100, 95, 90, 85, 80, 70, 60, 50, and 30%.

4. A series of dimethylbenzene–ethanol solution, with dimethylbenzene:ethanol = 100:0; 75:25; 50:50; 25:75, v/v. The 75:25 solution is saturated with safranin.

5. Paraplast plus (Sigma, P3683).

6. A slide warmer and a microtome.

7. Kimwipes (Kimberly-Clark).

8. Poly-Prep Slides (Sigma, P0425).

2.4. Solutions for the Hybridization and Color Visualization

1. The 20× SSC solution: 3 M NaCl, 0.3 M sodium citrate, pH 7.0.
2. The 2× SSC solution is made by 1:10 dilution of 20× SSC.
3. The 0.2× SSC solution is made by 1:10 dilution of 2× SSC.
4. 1 M Tris–HCl, pH 8.0.
5. Tris–EDTA solution: 100 mM Tris–HCl (pH 8.0), 50 mM EDTA. This solution can be made by 1:10 dilution of the above 1 M Tris–HCl (pH 8.0) and 0.5 M EDTA-Na.
6. Proteinase K stock: Dissolve 20 mg of proteinase K (MERCK, 1245680100) in 1 ml of water, and keep the solution at −20°C.
7. Proteinase K solution: Add 20 µl of protease K stock to 400 ml of Tris–EDTA solution.
8. 10× PBS: 1.3 M NaCl, 70 mM Na_2HPO_4, 30 mM NaH_2PO_4. Adjust pH to 7.0.
9. 1× PBS: 1:10 dilution of 10× PBS.
10. 10× glycine solution: Dissolve 2 g of glycine in 100 ml of 10× PBS. This solution is stable at 4°C for at least 1 month.
11. Glycine solution: Make a 1:10 dilution of the 10× glycine solution. Prepare 400 ml of glycine solution for 20 slides.
12. 4% Paraformaldehyde: Adjust pH of 1× PBS to 11.0 with NaOH, heat the solution to 60–70°C through a microwave oven, and add paraformaldehyde to a final concentration of 4%. Cool down the solution to 4°C on ice, and then adjust pH to 7.0 with HCl (see Note 1). Prepare 400 ml of 4% paraformaldehyde for 20 slides.
13. Acetic anhydride solution: Add 9 ml of triethanolamine to 600 ml of water, adjust pH to 8.0 by adding about 3 ml of HCl, and then add 3 ml of acetic anhydride (see Note 2).
14. Sodium phosphate solution: 1 M Na_3PO_4. Adjust pH to 6.8 with HCl.
15. 10× in situ salts: 3 M NaCl, 100 mM Tris–HCl (pH 8.0), 100 mM sodium phosphate (pH 6.8), 50 mM EDTA. Store at −20°C.
16. 50% dextran sulfate: Dissolve 5 g of dextran sulfate (Amresco 0198) in 10 ml of water. Store at −20°C.
17. tRNA stock solution: Dissolve 100 mg of tRNA (Sigma R4251) in 1 ml of water. Store at −20°C.
18. Hybridization solution A (a total of 1,600 µl for 20 slides): 200 µl of 10× in situ salts, 800 µl of deionized formamide (Sigma, F9037), 400 µl of 50% dextran sulfate, 40 µl of 50× Denhardt's (Sigma, D2532), 20 µl of tRNA stock solution, 140 µl of water.

19. Hybridization solution B (a total of 400 μl for 20 slides): 20 μl of probe, 180 μl of water, 200 μl of deionized formamide (see Note 3).
20. NTE solution (a total of 1.5 L is required for one experiment): 0.5 M NaCl, 10 mM Tris–HCl (pH 8.0), 1 mM EDTA.
21. Tris/NaCl solution (a total of 1.5 L is required for one experiment): 100 mM Tris–HCl (pH 7.5), 150 mM NaCl.
22. RNase stock: Dissolve 10 mg of RNase A (Sigma R4875) in 1 ml of 1:10 diluted Tris/NaCl solution, boil the solution for 15 min, and then store at –20°C.
23. RNase solution (a total of 400 ml for 20 slides): Make a 1:500 dilution of RNase stock using the NTE solution.
24. Blocking reagent: Dissolve 3 g of blocking reagent (Roche 11096176001) to 300 ml of Tris/NaCl solution.
25. BSA/Tris/Triton solution: 1% BSA; 0.3% Triton X-100, 100 mM Tris–HCl (pH 7.5), 150 mM NaCl. Dissolve 12 g of BSA in 1,200 ml of Tris/NaCl solution, and add 3.6 ml of Triton X-100 to this solution.
26. The anti-DIG antibody (Roche 11093274910).
27. Tris 9.5/NaCl/$MgCl_2$ solution: 100 mM Tris–HCl (pH 9.5), 100 mM NaCl, 50 mM $MgCl_2$.
28. NBT/BCIP solution (a total of 4 ml for 20 slides): Make a 1:50 dilution of NBT/BCIP (Roche 11681451001) with the Tris 9.5/NaCl/$MgCl_2$ solution.
29. Neutral balsam.

3. Methods

3.1. Probe Preparation

A cocktail is in an Eppendorf tube containing 5 μl of 20 pmol/μl probe, 5 μl of water, 4 μl of 5× reaction buffer, 4 μl of $CoCl_2$ solution, 1 μl of DIG-ddUTP solution, and 1 μl of terminal transferase. The contents are mixed well and pop spun for 1 s (see Note 4). The mixture is incubated at 37°C for 15 min, placed on ice, and 2 μl of 0.2 M EDTA (pH 8.0) is added to the mixture (see Note 5).

3.2. Sample Embedding

Fresh plant samples are placed into the FAA solution in a small glass bottle on ice, and subjected to a vacuum to remove free air in the plant tissues. The vacuum is usually terminated when the FAA solution starts to bubble. The plant specimens are kept under vacuum for approximately 20 min at 4°C (see Note 6) before the vacuum is slowly released. After vacuum treatment, the samples usually sink to the bottom of the glass bottle (see Note 7). The

FAA solution is changed once, and the samples are kept at 4°C overnight.

The plant samples are then dehydrated through a graded alcohol series (50, 60, 70, 85, 95, and 100% of ethanol for 60 min each; see Note 8), and the ethanol is changed with fresh 100% ethanol twice. The plant samples are treated with another graded series of mixed dimethylbenzene:ethanol, with a ratio of 25:75, 50:50, and 75:25 (v/v), for 30 min each.

The samples are then transferred into 100% dimethylbenzene for 60 min followed by a change of 100% dimethylbenzene solution, also for 60 min. A 1/4 volume of Paraplast Plus chips are added to the plant samples, which are left at room temperature overnight (see Note 9). The samples are then incubated at 42°C until the Paraplast Plus chips melt completely. Another 1/4 volume of Paraplast Plus chips are added to the samples, and incubation is continued at 60°C overnight. The old wax in the sample bottles is replaced with freshly melted Paraplast Plus two times, each for 3–4 days (see Note 10).

The samples are placed in paper boats containing 65°C freshly melted Paraplast Plus. After cooling, the embedded samples are kept at 4°C. The samples can be stored like this for at least 1 year.

3.3. Sectioning

The embedded samples are sectioned into 9-μm slices using a rotary microtome (see Note 11). Poly-Prep Slides are prewarmed at 42°C on a slide warmer. 1.5 ml of water is added onto the surface of a slide, and the ribbon of sample sections is arranged on the surface of water. Extra water is absorbed with a Kimwipe 1–2 min after the slide is covered with the appropriate number of ribbons flattened on the slide (see Note 12). The sections are kept at 42°C for 2 days.

3.4. Hybridization and Color Reaction

Day 1

1. Rehydration

   The slides are placed in a rack (slide holder for 20 slides) and soaked in 100% dimethylbenzene 10 min twice, and then in 100% ethanol for 2 min twice. The samples are rehydrated by successively soaking the slide rack in 95, 90, 80, 60, and 30% ethanol for 2 min each, in water for 2 min, and in 2× SSC for 20 min.

2. Protein digestion

   Protein in the samples is digested by placing the slide rack in protein K solution at 37°C for 30 min (see Note 13).

3. Refixation

   The slides are then refixed by incubating the slide rack in glycine solution for 2 min, in 1× PBS for 2×2 min, in 4% paraformaldehyde for 10 min, and then in 1× PBS for 2×5 min.

4. Acetic anhydride treatment

    The slides are treated with acetic anhydride by incubating the slide rack in acetic anhydride solution with gentle stirring for 10 min, and then in 1× PBS for 2×5 min.

5. Dehydration

    The samples are dehydrated by successively transferring the slide rack to a graded alcohol series of 30, 60, 80, 90, and 95% ethanol, each for 30 s, and changing to 100% ethanol for 2×30 s. All the slides from the slide rack are then placed horizontally in a plastic box, which contains 100% ethanol at the bottom, for 3–4 h at 4°C. Care is taken to avoid contacting the slides with the ethanol.

6. Hybridization

    Hybridization solution A is mixed with hybridization solution B in a 4:1 ratio, v/v. Bubbles in the mixed hybridization solution are removed by centrifugation. 100 µl of hybridization solution is added to the center of each slide, and the slides are covered with a piece of Parafilm to distribute the solution uniformly. The slides are placed in a plastic box containing 2×SSC and 50% formamide at the bottom (see Note 14). Care is taken to avoid contacting the slides with the formamide/SSC solution. The plastic box is incubated at 60°C overnight.

Day 2

7. Wash

    The next day, the slides are placed back in the slide rack and washed by soaking the rack in 0.2× SSC at 62°C for 60 min twice, in NTE solution at 37°C for 5 min twice, in RNAse solution at 37°C for 30 min, in NTE at 37°C for 5 min twice, in 0.2×SSC at 62°C for 60 min, and in 1×PBS at room temperature for 5 min (see Note 15).

8. Blocking reagent treatment and antibody reaction

    The slide rack is soaked in blocking reagent with gentle shaking for 45 min, and then transferred into BSA/Tris/Triton solution for another 45 min, with shaking. The anti-DIG antibody is diluted (1:1,250) with BSA/Tris/Triton solution, and 150 µl of the diluted DIG antibody is added to each slide. The slides are covered with Parafilm, placed in a plastic box containing BSA/Tris/Triton solution at the bottom, and the box is kept at room temperature for 2 h (see Note 14).

9. Color visualization

    The slides are incubated with gentle shaking in BSA/Tris/Triton solution at room temperature for 4×15 min. The BSA/Tris/Triton solution is then replaced with Tris 9.5/NaCl/$MgCl_2$ solution, and gently shaken for 10 min. The slides are

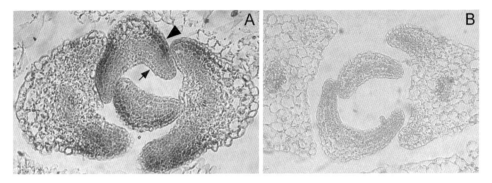

Fig. 3. In situ hybridization to analyze the distribution pattern of miR165. (**a**) The miR165 antisense LNA probe detected miR165 highly concentrated in the abaxial domain in wild-type leaf primordia. *An arrow* indicates the adaxial leaf side, while an *arrowhead* points to the abaxial side. (**b**) The miR165 sense LNA probe detects no signal.

briefly washed with fresh Tris 9.5/NaCl/MgCl$_2$ solution once, and then 200 µl of NBT/BCIP solution is added to each slide. The slides are covered with Parafilm and placed in a plastic box with water in the bottom. The plastic box is sealed with Parafilm, covered with foil to avoid light, and placed at 4°C (see Note 16).

10. Checking the results of the color reaction

    It is necessary to check the results of the color reaction under a microscope from time to time. Here, we use the distribution of *Arabidopsis* miR165 in the leaf primordium as an example. Compared with the result using the sense probe, LNA-miR165, hybridization with the antisense probe resulted in signal accumulation in the abaxial side of the leaf primordium (Fig. 3).

3.5. Slides' Preservation

For preservation, slides are dehydrated through successive incubations in 30, 50, 60, 70, 85, 95, and 100% ethanol, each for 5 s, and changed to 100% ethanol once. The slides are placed in 100% dimethylbenzene for 10 min twice, and then mounted by adding four drops of neutral balsam to the center of a slide, followed by covering the slide with a microscope coverslip to allow uniform spreading of the neutral balsam. The slides are kept at room temperature overnight.

4. Notes

1. Paraformaldehyde is toxic. Operation with paraformaldelyde must be carried out inside a fume cupboard. In addition, 4% paraformaldehyde should be prepared freshly before use.
2. Freshly add acetic anhydride before use.

3. Calculate the number of each group of slides hybridized using the same probe, and then prepare the corresponding volume of hybridization solution B. Before adding the probe into the rest of hybridization solution B, the probe solution should be treated for 2 min at 80°C, and placed on ice immediately for denaturation. After completely cooling down, solution B is spun down and kept on ice.

4. For defrosting, the reaction buffer, $CoCl_2$, and DIG-ddUTP should be kept at 4°C before use.

5. Usually, 0.5–1 µl of the labeled probe is used for each slide, but for a newly made probe, we suggest performing a preexperiment to detect the hybridization efficiency of a graded series of probe concentrations. Use 0.5, 1, 2, and 4 µl for in situ hybridization to find the suitable volume for each slide.

6. Increase and release the pressure slowly. Young seedlings should be subjected to a vacuum for 20 min, while siliques need 40–80 min.

7. If the samples do not sink to the bottom, repeat the vacuum procedure.

8. Gently shake the glass bottle from time to time to improve the dehydration efficiency. Plant samples can be stored in 70% ethanol for at least 6 months.

9. All operations with dimethylbenzene should be carried out in a fume cupboard.

10. More frequent change of Paraplast Plus can shorten the time from 4 days to 2 to 3 days.

11. Use chloroform to wipe the blade, the knife of the rotary microtome, and the platform of microscope to avoid the contamination with RNase.

12. Check slides under the microscope to identify ideal sections during sectioning.

13. The proteinase K solution should be pre-warmed to 37°C before use.

14. When covering slides with Parafilm, avoid making bubbles, which will severely influence hybridization signals. Use Parafilm to seal the box to retain moisture.

15. All solutions should be pre-warmed to the corresponding temperature before use. Shake gently in each step to improve the efficiency. After washing, the slides can be kept at 4°C.

16. Check the slides after 10 h from time to time under the microscope to get clear signals. Add additional NBT/BCIP solution to the slides to continue the color reaction after microscope observation.

Acknowledgments

We are grateful to H. Wang for discussion with the manuscript. We thank Dr. Y. Eshed and Dr. I. Pekker for sharing experience on LNA probes. This work was supported by grants from the Chief Scientist Program of Shanghai Institutes for Biological Sciences, the Chinese National Scientific Foundation 30630041, and Chinese Academy of Sciences (KSCX2-YW-N-057).

References

1. Bartel DP (2004) MicroRNAs: genomics, biogenesis, mechanism, and function. Cell 116: 281–297
2. Wienholds E, Plasterk RH (2005) MicroRNA function in animal development. FEBS Lett 579:5911–5922
3. Wu L, Zhou H, Zhang Q, Zhang J, Ni F, Liu C, Qi Y (2010) DNA methylation mediated by a microRNA pathway. Mol Cell 38:465–475
4. Chellappan P, Xia J, Zhou X, Gao S, Zhang X, Coutino G, Vazquez F, Zhang W, Jin H (2010) siRNAs from miRNA sites mediate DNA methylation of target genes. Nucleic Acids Res 38:6883–6894
5. Reinhart BJ, Slack FJ, Basson M, Pasquinelli AE, Bettinger JC, Rougvie AE, Horvitz HR, Ruvkun G (2000) The 21-nucleotide let-7 RNA regulates developmental timing in *Caenorhabditis elegans*. Nature 403:901–906
6. Aukerman MJ, Sakai H (2003) Regulation of flowering time and floral organ identity by a MicroRNA and its APETALA2-like target genes. Plant Cell 15:2730–2741
7. Mallory AC, Bartel DP, Bartel B (2005) MicroRNA-directed regulation of *Arabidopsis* AUXIN RESPONSE FACTOR17 is essential for proper development and modulates expression of early auxin response genes. Plant Cell 17:1360–1375
8. Yang L, Liu Z, Lu F, Dong A, Huang H (2006) SERRATE is a novel nuclear regulator in primary microRNA processing in Arabidopsis. Plant J 47:841–850
9. Diederichs S, Haber DA (2007) Dual role for argonautes in microRNA processing and post-transcriptional regulation of microRNA expression. Cell 131:1097–1108
10. Parizotto EA, Dunoyer P, Rahm N, Himber C, Voinnet O (2004) In vivo investigation of the transcription, processing, endonucleolytic activity, and functional relevance of the spatial distribution of a plant miRNA. Genes Dev 18:2237–2242
11. Wang JW, Wang LJ, Mao YB, Cai WJ, Xue HW, Chen XY (2005) Control of root cap formation by MicroRNA-targeted auxin response factors in Arabidopsis. Plant Cell 17:2204–2216
12. Jung JH, Park CM (2007) MIR166/165 genes exhibit dynamic expression patterns in regulating shoot apical meristem and floral development in Arabidopsis. Planta 225:1327–1338
13. Raman S, Greb T, Peaucelle A, Blein T, Laufs P, Theres K (2008) Interplay of miR164, CUP-SHAPED COTYLEDON genes and LATERAL SUPPRESSOR controls axillary meristem formation in *Arabidopsis thaliana*. Plant J 55:65–76
14. Gutierrez L, Bussell JD, Pacurar DI, Schwambach J, Pacurar M, Bellini C (2009) Phenotypic plasticity of adventitious rooting in Arabidopsis is controlled by complex regulation of AUXIN RESPONSE FACTOR transcripts and microRNA abundance. Plant Cell 21:3119–3132
15. Yao X, Wang H, Li H, Yuan Z, Li F, Yang L, Huang H (2009) Two types of cis-acting elements control the abaxial epidermis-specific transcription of the MIR165a and MIR166a genes. FEBS Lett 583:3711–3717
16. Yoon EK, Yang JH, Lim J, Kim SH, Kim SK, Lee WS (2010) Auxin regulation of the microRNA390-dependent transacting small interfering RNA pathway in Arabidopsis lateral root development. Nucleic Acids Res 38:1382–1391
17. Rodriguez JB, Marquez VE, Nicklaus MC, Mitsuya H, Barchi JJ Jr (1994) Conformationally locked nucleoside analogues. Synthesis of dideoxycarbocyclic nucleoside analogues structurally related to neplanocin C. J Med Chem 37:3389–3399

18. Braasch DA, Corey DR (2001) Locked nucleic acid (LNA): fine-tuning the recognition of DNA and RNA. Chem Biol 8:1–7
19. Wengel J, Petersen M, Nielsen KE, Jensen GA, Hakansson AE, Kumar R, Sorensen MD, Rajwanshi VK, Bryld T, Jacobsen JP (2001) LNA (locked nucleic acid) and the diastereoisomeric alpha-L-LNA: conformational tuning and high-affinity recognition of DNA/RNA targets. Nucleosides Nucleotides Nucleic Acids 20:389–396
20. Hakansson AE, Wengel J (2001) The adenine derivative of alpha-L-LNA (alpha-L-ribo configured locked nucleic acid): synthesis and high-affinity hybridization towards DNA, RNA, LNA and alpha-L-LNA complementary sequences. Bioorg Med Chem Lett 11:935–938
21. Randazzo A, Esposito V, Ohlenschlager O, Ramachandran R, Virgilio A, Mayol L (2005) Structural studies on LNA quadruplexes. Nucleosides Nucleotides Nucleic Acids 24:795–800
22. Valoczi A, Hornyik C, Varga N, Burgyan J, Kauppinen S, Havelda Z (2004) Sensitive and specific detection of microRNAs by northern blot analysis using LNA-modified oligonucleotide probes. Nucleic Acids Res 32:e175
23. Wienholds E, Kloosterman WP, Miska E, Alvarez-Saavedra E, Berezikov E, de Bruijn E, Horvitz HR, Kauppinen S, Plasterk RH (2005) MicroRNA expression in zebrafish embryonic development. Science 309:310–311
24. Kloosterman WP, Wienholds E, de Bruijn E, Kauppinen S, Plasterk RH (2006) In situ detection of miRNAs in animal embryos using LNA-modified oligonucleotide probes. Nat Methods 3:27–29
25. Valoczi A, Varallyay E, Kauppinen S, Burgyan J, Havelda Z (2006) Spatio-temporal accumulation of microRNAs is highly coordinated in developing plant tissues. Plant J 47:140–151
26. Long JA, Moan EI, Medford JI, Barton MK (1996) A member of the *KNOTTED* class of homeodomain proteins encoded by the STM gene of *Arabidopsis*. Nature 379:66–69
27. You Y, Moreira BG, Behlke MA, Owczarzy R (2006) Design of LNA probes that improve mismatch discrimination. Nucleic Acids Res 34:e60

Chapter 12

Small RNA Isolation and Library Construction for Expression Profiling of Small RNAs from *Neurospora* and *Fusarium* Using Illumina High-Throughput Deep Sequencing

Gyungsoon Park and Katherine A. Borkovich

Abstract

Due to crucial roles in gene regulation, noncoding small RNAs (smRNAs) of 20–30 nucleotides (nt) have been intensively studied in mammals and plants, and are known to be implicated in significant diseases and metabolic disorders. Elucidation of biogenesis mechanisms and functional characterization of smRNAs are often achieved using tools, such as separation of small-sized RNA and high-throughput sequencing. Although RNA interference pathways such as quelling and meiotic silencing have been well described in *Neurospora crassa*, knowledge of smRNAs in filamentous fungi is still limited compared to other eukaryotes. As a prerequisite for study, isolation and sequence analysis of smRNAs are necessary. We developed a protocol for isolation and library construction of smRNAs of 20–30 nt for Solexa sequencing in two filamentous fungi, *N. crassa* and *Fusarium oxysporum f.sp. lycopersici*. Using 200–300 μg total RNA, smRNA was isolated by size fractionation, ligated with adapters, and amplified by RT-PCR for Solexa sequencing. Sequence analysis of several cDNA clones showed that the cloned smRNAs were not tRNAs and rRNAs and were fungal genome specific.

Key words: Small RNA isolation, *Neurospora crassa*, *Fusarium oxysporum f.sp. lycopersici*, Filamentous fungi, Library construction

1. Introduction

Noncoding small RNAs (smRNAs) of 20–30 nt with regulatory functions have been frequently demonstrated in eukaryotes (1, 2). Small RNAs include micro RNAs (miRNAs) and small interfering RNAs (siRNAs). The difference between these two derives from their mode of generation. miRNAs are produced from endogenous transcripts with a local hairpin structure, whereas siRNAs arise from double-strand RNA transcripts (3). Small RNAs regulate gene expression posttranscriptionally or post-translationally. They pair with complementary sequences in their target mRNAs and induce mRNA degradation, or with the 3′ UTR of mRNAs and interfere with translation (4, 5).

Fungal smRNAs have been reported in a few species. Noncoding RNAs of ~400–700 nucleotides (nt) have been isolated and shown to regulate transcripts at the posttranscriptional or post-translational level in the yeast *Saccharomyces cerevisiae* (6, 7). In the filamentous fungi, *Neurospora crassa* has been reported to produce microRNA-like small RNAs (milRNAs) and dicer-independent siRNAs (disiRNAs) (8, 9). Recently, in the filamentous fungal pathogen *Magnaporthe grisea*, methylguanosine-capped and polyadenylated smRNAs were discovered through a series of procedures, such as RNA selection, full-length cDNA cloning, and 454 sequencing (10). Although an increasing number of studies are reporting the discovery and functional characterization of smRNAs, biogeneration, genomic origin, and induction of smRNAs still remain to be clarified. Population analysis of smRNAs is crucial for revealing regulatory mechanisms, and construction and sequencing of smRNA libraries is a necessary first step toward this goal.

In this chapter, we describe a procedure for isolating smRNAs of 20–30 nt and constructing libraries for Solexa sequence analysis in two filamentous fungi, *N. crassa* and *Fusarium oxysporum*. *N. crassa* is a nonpathogenic filamentous fungus with ~10,000 predicted genes and serves as a model for species that are important pathogens of plants and animals (11). *F. oxysporum* is a soil-borne pathogen that infects various vegetables, fruits, and ornamental plants. This fungus is classified into many formae speciales, depending on the natural hosts infected. *F. oxysporum f.sp. lycopersici* causes vascular wilt in tomato and has ~18,000 predicted genes (12).

2. Materials

2.1. Growth of Fungal Cultures

1. *N. crassa* strain FGSC 4200 (wild type, *mat a*).
2. *F. oxysporum f.sp. lycopersici* (USDA 34936).
3. 50× Vogel's Minimal (VM) Medium Salts (1 L): 126.8 g sodium citrate dihydrate, 250 g potassium phosphate (monobasic), 100 g ammonium nitrate, 10 g magnesium sulfate.7H$_2$O, 5 g calcium chloride.2H$_2$O, 5 ml biotin solution, 5 ml trace elements solution, 5 ml chloroform as preservative.
4. Biotin solution (filter sterilize): 5 mg biotin/100 ml 50% (v/v) ethanol.
5. Trace elements (filter sterilize): In 100 ml, 5 g citric acid monohydrate, 5 g zinc sulfate.7H$_2$O, 1 g ferrous ammonium sulfate.6H$_2$O, 0.25 g cupric sulfate.5H$_2$O, 0.05 g manganese sulfate.H$_2$O, 0.05 g boric acid, 0.05 g sodium molybdate.2H$_2$O.

6. VM: 1× VM salts, 1.5% sucrose, 1% agar (for agar medium).
7. VM agar flasks: Foam-stoppered 125-ml Erlenmeyer glass flasks containing 25 ml 1× VM agar medium (autoclave before using).
8. Masslin shop towel (Chicopee, Benson, NC).
9. Incubator with light (25–30°C).
10. Clinical centrifuge.
11. Hemacytometer.

2.2. Total RNA Extraction

1. Extraction buffer: 100 mM Tris–HCl (pH 8.0), 100 mM lithium chloride, 10 mM EDTA, 1% SDS.
2. Diethylpyrocarbonate (DEPC)-treated ddH$_2$O (DEPC-H$_2$O).
3. 70% Ethanol in DEPC-H$_2$O (v/v).
4. 4 M Lithium chloride.
5. 24:1 (v/v) Chloroform–isoamyl alcohol.
6. Water-saturated phenol, pH 4.5.
7. Mortar and pestle.
8. Liquid nitrogen.
9. 45-ml Oakridge centrifuge tubes and refrigerated high-speed centrifuge.

2.3. Small RNA Library Construction

1. Vertical midi gel electrophoresis system (16×20 cm); Protean II xi vertical electrophoresis cells (Bio-Rad, Hercules, CA).
2. 40% acrylamide:bis-acrylamide (29:1).
3. Urea.
4. 5× TBE: 54 g Tris, 27.5 g boric acid, 20 ml 0.5 M EDTA (pH 8.0); bring to 1 L final volume.
5. Tetramethylethylenediamine (TEMED).
6. Ammonium persulfate (APS).
7. Ethidium bromide.
8. 2× gel loading buffer (95% formamide, 18 mM EDTA (pH 8.0), 0.025% SDS, 0.01 g xylene cyanol, 0.01 g bromophenol blue).
9. 0.3 M NaCl.
10. Spin-X Cellulose Acetate filter (Fisher Scientific, Pittsburgh, PA).
11. DEPC-H$_2$O.
12. Glycogen (5 mg/ml) (Fisher Scientific, Pittsburgh, PA).
13. Room-temperature 75% ethanol.
14. 10 bp DNA ladder (Fermentas, Glen Burnie, MA).
15. 200-μl RNase-free PCR tubes.

16. T4 RNA ligase (10 U/μl; Promega, Madison, WI).
17. RNaseOut™ (40 U/μl; Invitrogen, Carlsbad, CA).
18. SuperScript™ II Reverse Transcription Kit (Invitrogen, Carlsbad, CA).
19. Phusion high fidelity DNA polymerase (New England Biolabs, Ipswich, MA).
20. Pellet Paint® NF Co-precipitant (Novagen, Gibbstown, NJ).
21. 3 M sodium acetate (pH 5.2).
22. Razor blade.
23. PCR thermocycler.
24. Platform rotator.
25. Resuspension buffer (10 mM Tris–HCl, pH 8.5).
26. 5′ Adapter; GUUCAGAGUUCUACAGUCCGACGAUC.
27. 3′ Adapter; pUCGUAUGCCGUCUUCUGCUUGidU.
28. 3′ Small RNA RT primer; CAAGCAGAAGACGGCATACGA.
29. 5′ Small RNA PCR Primer; AATGATACGGCGACCACCGACAGGTTCAGAGTTCTACAGTCCGA.
30. pGEMT vector systems (Promega, Madison, WI).

3. Methods

3.1. Growth of Fungal Cultures

1. Inoculate foam-stoppered VM agar in 125-ml flask with fungal spores or a mycelial fragment. Incubate at 30°C in the dark for 2 days and then at 25°C in the light for 5 days (for *N. crassa*) or at 28°C in the light for 10 days (for *F. oxysporum*).
2. Add 50 ml of sterile water into the flasks and vortex to disperse spores.
3. Filter suspension through two layers of Masslin shop towel and collect the conidial suspension in a 50-ml Falcon tube.
4. Spin down conidia at maximum speed for 5 min in a clinical centrifuge.
5. After removing the supernatant, add 30 ml of sterile water and vortex.
6. Spin down the conidia as in step 4.
7. After removing the supernatant, resuspend the conidia in 1 ml of sterile water.
8. Determine the conidial concentration using a hemacytometer.
9. Inoculate conidial suspension into 1 L of VM liquid medium at a concentration of 5×10^6 conidia/ml.

10. Incubate the flask with shaking for 24 h at 28°C.
11. Harvest the fungal mycelia (cell pad) by filtering the culture through two layers of Masslin shop towels.
12. Gently squeeze out excess water and store the cell pad at −80°C until RNA extraction.

3.2. Total RNA Extraction Using Hot Phenol Method (13)

1. Heat phenol (pH 4.5) and extraction buffer to 80°C for at least 10 min using a water bath.
2. Freeze the fungal mycelium in liquid nitrogen and grind to a fine powder in a mortar and pestle (grind three to four times).
3. Transfer the ground powder to a 45-ml Oakridge centrifuge tube using a funnel and keep the tube on ice until liquid nitrogen has evaporated.
4. Add 7 ml of hot phenol and 7 ml of hot extraction buffer to the powder, and vortex for 30 s.
5. Add 7 ml of chloroform/isoamyl alcohol (24:1) to the solution and vortex for 30 s.
6. Keep the tube on ice while processing other samples, if needed (six samples maximum).
7. Centrifuge the tube at $12,000 \times g$ for 15 min at 4°C.
8. Transfer the aqueous phase to a clean 45-ml Oakridge centrifuge tube.
9. Repeat steps 5–8 twice.
10. Add an equal volume of 4 M lithium chloride to the solution in the tube and precipitate RNA on ice overnight at 4°C (16–18 h).
11. Centrifuge the tube at $12,000 \times g$ for 30 min at 4°C.
12. Pour off the supernatant and rinse the pellet with 20 ml of 70% ethanol.
13. Pour off the 70% ethanol and air dry the pellet for about 15 min at room temperature or on ice.
14. Resuspend the pellet in 100–200 µl of DEPC-H_2O (yield is approximately 200–400 µg total RNA).

3.3. Small RNA Library Construction

3.3.1. Purification of 20–30-nt RNAs from 200 to 400 µg of Total RNA

1. Prepare 17% polyacrylamide–urea gel (for a 16×20-cm gel) in an RNase-free 200-ml bottle: Mix 6 ml of water, 4 ml of 5×TBE, 17 ml of acrylamide:bis-acrylamide (29:1), and 17 g of urea. Heat mixture at 42°C to dissolve components (volume will reach 40 ml when dissolved).
2. When cool, add 70 µl of 25% APS and 20 µl of TEMED, and mix.

3. Carefully pour the gel solution between glass gel plates (PROTEAN II xi Cell) using a 25-ml pipet (avoid bubbles) and then insert the comb.
4. After polymerization (~20 min), place the gel plates in central cooling core with gaskets and pour running buffer (1× TBE) into upper buffer dam.
5. Carefully remove the comb and thoroughly rinse wells three times using syringe to eliminate excess urea.
6. Place the assembly in the electrophoresis chamber.
7. Pre-run the gel for 15–30 min at 200 V.
8. While the gel is pre-running, mix total RNA with an equal volume of 2× gel loading buffer and heat at 65°C for 5 min.
9. Mix 10 µl of 10 bp DNA ladder with 10 µL of 2× gel loading buffer in another microfuge tube. Do not heat this sample prior to loading on the gel.
10. After washing wells again, load samples and 10 bp ladder.
11. Run the gel at 250 V for about 4–5 h (until purple dye is 4 cm above the bottom of gel).
12. Stain the gel in ethidium bromide (0.5 mg/L)/1× TBE for 15 min.
13. Excise the area of the gel corresponding to 20–30 nt with a clean razor blade and chop into small slices on an RNase-free glass plate.
14. Transfer the chopped gel band to an RNase-free microcentrifuge tube.
15. Add 100 µl of 0.3 M NaCl and crush the gel slices using an RNase-free plastic rod.
16. Add 900 µl (or enough to barely cover the crushed gel) of 0.3 M NaCl to the tube (seal the tube with Parafilm) and elute the DNA by rotating the tube gently at 4°C overnight or at room temperature for 4 h.
17. Transfer the liquid to a Spin-X Cellulose Acetate filter and centrifuge in a microcentrifuge for 2 min at maximum speed.
18. Transfer the liquid to a new tube, and add 1/100 volume of glycogen (5 mg/ml) and an equal volume of isopropanol.
19. Precipitate the RNA at –80°C for 2–3 h.
20. Centrifuge at $12,000 \times g$ for 30 min in a 4°C microcentrifuge.
21. Carefully remove the supernatant and wash the pellet (by pipeting in and out once very slowly and smoothly) with 500 µl of room-temperature 75% ethanol.
22. Air dry the RNA pellet (about 2 min) and then dissolve the RNA in 5.7 µl of DEPC-treated water.

3.3.2. 5′ Adaptor Ligation and Purification

1. Set up the 5′Adaptor ligation reaction in a 200-μl RNase-free PCR tube: 5.7 μl Purified smRNA, 1.3 μl 5′ RNA adaptor (5 μM), 1 μl 10× RNA ligation buffer, 1 μl T4 RNA ligase (10 U/μl), and 1 μl RNaseOut™ (40 U/μl).
2. Incubate at 20°C for 6 h in a thermal cycler and then hold at 4°C overnight.
3. Stop reaction by adding 10 μl 2× gel loading buffer.
4. Heat the ligated sample/loading buffer mixture at 65°C for 5 min.
5. Combine 10 μl of 10 bp DNA ladder with 10 μl of 2× gel loading buffer in another microcentrifuge tube.
6. Prepare 14% TBE-urea gel as described above.
7. After washing wells and pre-running the gel, load the ligated samples and ladder on the gel.
8. Run the gel at 250 V until the bromophenol blue dye (light blue) migrates to 5 cm above the bottom of the gel (dark-blue xylene cyanol will have run off the end of the gel).
9. Stain the gel as in Subheading 3.3.1, step 12, above.
10. Cut out the gel band corresponding to 40–60 nt with a clean razor blade.
11. Chop the gel and place in a microcentrifuge tube. Then, add 200 μl of 0.3 M NaCl and crush the gel using an RNase-free plastic rod.
12. Add an additional 300 μl or more of 0.3 M NaCl to the tube and elute the RNA by rotating the tube gently at 4°C overnight or at room temperature for 4 h.
13. Centrifuge the tube and collect the liquid into a new tube using a Spin-X Cellulose Acetate filter (as described above).
14. Add 1/100 volume of glycogen and an equal volume of isopropanol to the sample and incubate at −80°C for 2–3 h.
15. Centrifuge the tube at 12,000×g for 30 min in a 4°C microcentrifuge.
16. Carefully remove the supernatant and wash the pellet as indicated in Subheading 3.3.1, step 21, above.
17. Allow the RNA pellet to air dry (about 2 min) and then dissolve the RNA in 6.4 μl of DEPC-treated water.

3.3.3. 3′ Adaptor Ligation and Purification

1. Set up the 3′Adaptor ligation reaction in a 200-μl RNase-free microfuge tube: 6.4 μl Purified 5′ ligation product, 0.6 μl 3′ RNA adaptor (10 μM), 1 μl 10× RNA ligation buffer, 1 μl T4 RNA ligase (10 U/μl), and 1 μl RNaseOut™ (40 U/μl).
2. Incubate the tube at 20°C for 6 h in a thermal cycler and then hold at 4°C overnight.

3. Follow steps 3–6 in Subheading 3.3.2, above.
4. After washing the wells and pre-running a 14% TBE-urea gel, load the ligated samples and ladder on the gel.
5. Run the gel at 250 V as in Subheading 3.3.2, step 8, above.
6. Stain the gel as in Subheading 3.3.1, step 12, above.
7. Cut out the region of the gel corresponding to 60–80 nt with a clean razor blade.
8. Chop the gel and place in a tube. Then, add 200 µl of 0.3 M NaCl and crush the gel using an RNase-free plastic rod.
9. Add 300 µl of 0.3 M NaCl to the tube and elute the RNA by rotating the tube gently at 4°C overnight or at room temperature for 4 h.
10. Centrifuge and collect the liquid to a new tube using Spin-X Cellulose Acetate filter (as described above).
11. Add 1/100 volume of glycogen and equal volume of isopropanol to the sample and incubate at –80°C for 2–3 h.
12. Centrifuge the tube at $12,000 \times g$ for 30 min at 4°C.
13. Carefully remove the supernatant and wash the pellet as in Subheading 3.3.1, step 21, above.
14. Allow the RNA pellet to air dry (about 2 min) and then dissolve the RNA in 4.5 µl of DEPC-treated water.

3.3.4. RT-PCR of Small RNAs Ligated with Adaptors

1. Set up a reverse transcription reaction in a 0.5-ml RNase-free microfuge tube: 4.5 µl Purified ligated RNA and 0.5 µl 3′ Small RNA RT primer (100 µM).
2. Heat to 65°C for 5 min. Centrifuge briefly at $12,000 \times g$ to bring solution to the bottom of the tube.
3. Add the following in order: 2.0 µl 5× First-strand buffer, 0.5 µl 12.5 mM dNTP mix, 1 µl 100 mM DTT, and 0.5 µL RNaseOut™ (40 U/µl).
4. Heat tube to 48°C for 3 min and then add 1.0 µl of SuperScript™ II RT (200 U/µl).
5. Incubate at 42°C for 1 h.
6. Set up a PCR reaction from the reverse transcription samples: 10 µl of RT reaction mix from step 5, 10 µl 5× Phusion™ HF Buffer, 0.5 µl 5′ Small RNA PCR Primer (25 µM), 0.5 µl 3′ Small RNA RT Primer (25 µM), 0.5 µl 12.5 mM dNTP mix, 0.5 µl Phusion™ DNA Polymerase, and 18 µl H_2O.
7. PCR thermocyler conditions, 98°C for 30 s, 15 cycles at 98°C for 10 s, 60°C for 30 s, and 72°C for 15 s; final extension at 72°C for 10 min.

3.3.5. Gel Purification of RT-PCR Product

1. Load 10 μl of 10-bp DNA marker into one well and PCR products into another well of 14% TBE PAGE gel (no urea): 22 ml Water, 4 ml 5× TBE, 14 ml acrylamide:bis-acrylamide (29:1), 70 μl 25% APS, and 20 μl TEMED.
2. Run gel at 200 V as in Subheading 3.3.2, step 8, above.
3. Stain the gel in ethidium bromide (0.5 mg/L)/1× TBE for 10 min.
4. Cut out the region corresponding to ~94–100 nt with a clean razor blade and transfer the gel band into a microcentrifuge tube.
5. Crush the gel band and add 500 μl of 1× NEB Restriction Enzyme Buffer 2 to the tube.
6. Elute the DNA by rotating the tube gently at 4°C overnight or at room temperature for 2 h.
7. Centrifuge the tube and collect the liquid to a new tube using a Spin-X Cellulose Acetate filter (as described above).
8. Add 1/10 vol. of 3 M sodium acetate, an equal volume of isopropanol, and 1 μl of Pellet Paint® NF Co-precipitant to the sample and incubate at −80°C for 2 h.
9. Centrifuge at $12,000 \times g$ for 30 min at 4°C.
10. Carefully remove the supernatant and wash the pellet as in Subheading 3.3.1, step 21, above.
11. Centrifuge the tube briefly at $12,000 \times g$ to pool the remaining ethanol in the bottom of the tube.
12. Allow the DNA pellet to air dry (about 2 min) and then resuspend in 12 μl of resuspension buffer (10 mM Tris–HCl, pH 8.5).
13. Take 1 μl and A-tail library clones following the manufacturer's recommendations (pGEM®-T and pGEM®-T Easy Vector Systems manual).
14. Perform TA cloning using pGEM®-T and pGEM®-T Easy Vector Systems and sequence about 20 clones to check the quality of library.

4. Notes

This protocol was designed for isolating small RNAs from fungal mycelia grown in liquid medium. The fungal mycelial mat should be completely dried and frozen at −80°C for several minutes after harvesting in order to facilitate efficient grinding. At least 200 μg of total RNA is needed to ensure isolation of an adequate quantity of small RNAs. During electrophoresis of small RNAs, the wells of polyacrylamide–urea gels should be thoroughly washed to prevent urea precipitation. Crushing the gel slices prior to elution greatly facilitates isolation of small RNAs.

Acknowledgments

This work was supported by a grant from the UCR-LANL Collaborative Program in Pathogen-Induced Plant Infectious Disease to Katherine Borkovich. Gyungsoon Park was partially supported by the National Research Foundation of Korea Grant, funded by the Korean Government (No.2010-0029418 20110014825).

References

1. Bartel DP (2004) MicroRNAs: genomics, biogenesis, mechanism, and function. Cell 116(2): 281–297
2. Yang TW, Xue LG, An LZ (2007) Functional diversity of miRNA in plants. Plant Science 172(3):423–432
3. Finnegan EJ, Matzke MA (2003) The small RNA world. J Cell Sci 116(Pt 23):4689–4693
4. Olsen PH, Ambros V (1999) The *lin-4* regulatory RNA controls developmental timing in Caenorhabditis elegans by blocking LIN-14 protein synthesis after the initiation of translation. Dev Biol 216(2):671–680
5. Hammond SM et al (2000) An RNA-directed nuclease mediates post-transcriptional gene silencing in Drosophila cells. Nature 404(6775): 293–296
6. Neil H et al (2009) Widespread bidirectional promoters are the major source of cryptic transcripts in yeast. Nature 457(7232):1038–1042
7. Xu Z et al (2009) Bidirectional promoters generate pervasive transcription in yeast. Nature 457(7232):1033–1037
8. Lee HC et al (2009) qiRNA is a new type of small interfering RNA induced by DNA damage. Nature 459(7244):274–277
9. Lee HC et al (2010) Diverse pathways generate microRNA-like RNAs and Dicer-independent small interfering RNAs in fungi. Mol Cell 38(6):803–814
10. Gowda M et al (2010) Genome-wide characterization of methylguanosine-capped and polyadenylated small RNAs in the rice blast fungus *Magnaporthe oryzae*. Nucleic Acids Res 38(21):7558–7569
11. Galagan JE et al (2003) The genome sequence of the filamentous fungus *Neurospora crassa*. Nature 422(6934):859–868
12. Ma LJ et al (2010) Comparative genomics reveals mobile pathogenicity chromosomes in Fusarium. Nature 464(7287):367–373
13. Verwoerd TC, Dekker BM, Hoekema A (1989) A small-scale procedure for the rapid isolation of plant RNAs. Nucleic Acids Res 17:2362

Chapter 13

Isolation and Profiling of Protein-Associated Small RNAs

Hongwei Zhao, Yifan Lii, Pei Zhu, and Hailing Jin

Abstract

Small RNAs are short noncoding RNAs with important regulatory roles in many cellular processes. Small RNAs are generated by DICER or DICER-like proteins and then incorporated into RNAi effector proteins ARGONAUTEs (AGOs) for silencing of their targets. In plants, small RNAs regulate host innate immunity against various pathogens, but their mode of action and associated protein factors that facilitate their function remain to be elucidated. Here, we describe an efficient method to isolate AGO-associated small RNAs from *Arabidopsis*. This protocol can be easily adapted for the isolation of any protein-associated small RNAs. We utilized immunoprecipitation tandem with deep sequencing to identify small RNAs with functions in plant innate immunity. Using this described protocol, we identified miR393* that plays a crucial role in plant antibacterial defense. The distinct roles played by individual AGO proteins were observed.

Key words: Immunoprecipitation, Argonaute, Protein-associated small RNAs, Plant innate immunity

1. Introduction

Small RNAs are noncoding RNAs that regulate gene expression in a sequence-specific manner. They are generally 20–30 nucleotides (nt) in length and can be categorized as microRNAs (miRNAs) or small interfering RNAs (siRNAs) according to their precursor structures and biogenesis pathways (1, 2). Small RNAs are involved in the regulation of many growth and developmental processes, as well as stress responses, diseases, and metabolic disorders (1–4).

Small RNAs are generated by DICER or DICER-like (DCL) proteins and then incorporated into effector complexes (RNA-induced silencing complex [RISC]) containing an ARGONAUTE (AGO) protein to guide silencing of their target RNAs (5).

Plants have various AGO proteins with distinct functions. In *Arabidopsis*, ten AGO proteins were identified (6). AGO1 binds mainly miRNAs, while AGO4 and AGO6 bind mainly 24-nt heterochromatic siRNAs. AGO2 is highly induced by bacterial or viral pathogens and regulates plant innate immunity (7, 8). *Arabidopsis* small RNAs load into corresponding AGO-containing RISC usually according to the identity of their first nucleotide (9, 10).

Most RNAs, including small RNAs, are regulated and function through RNA-binding proteins. Small RNAs are loaded into AGO proteins to induce gene silencing of their targets. To unveil the biological roles of small RNAs, it is necessary to investigate the expression profiles of AGO-associated small RNAs (9–11). AGO immunoprecipitation (IP) (pull-down) tandem with deep sequencing can provide a snapshot of in vivo small RNA populations in different AGO complexes (9, 10). Such information is indispensable for dissecting the biological roles and functionality of small RNAs. In this chapter, we describe an efficient protocol for isolating and profiling AGO-associated small RNAs from *Arabidopsis*. We have isolated AGO1- and AGO2-associated small RNAs in healthy (mock treated) and bacterial pathogen (*Pseudomonas syringae*)-challenged *Arabidopsis* samples (7). The deep-sequencing results allowed us to identify miR393* as a functional miRNA specifically enriched in AGO2 from bacterial pathogen-challenged plants, revealing its important role in plant innate immunity. We demonstrate that miR393* and miR393 specifically load into AGO2 and AGO1, respectively (7).

2. Materials

2.1. Bacterial Pathogen Inoculation

A. *Arabidopsis* Columbia-0 carrying an HA-tagged GUS gene, used as a control.

B. *Arabidopsis* Columbia-0 plants carrying an HA-tagged AGO2 (*AGO2::3HA:AGO2*).

C. *Pseudomonas syringae* pv. *tomato* DC3000 carrying *avrRpt2* (*Pst* [*avrRpt2*]).

D. Pseudomonas Agar F (PAF) medium: 19 g BD Difco™ PAF powder, 5 ml glycerol. Make up to 500 ml with distilled water and autoclave. Cool to 50°C and then add 50 μg/ml rifampicin and 50 μg/ml kanamycin before pouring plates.

E. 10 mM $MgCl_2$: In sterile distilled water.

2.2. Protein Pull-Down

A. IP extraction buffer (pH 7.5): 20 mM Tris–HCl, 300 mM NaCl, 5 mM $MgCl_2$, 0.5% (v/v) NP40, 5 mM DTT (added prior to use), 1 tablet/50 ml protease inhibitor (added prior to use), make up to 50 ml with DEPC-treated water.

B. IP washing buffer (pH 7.5): 20 mM Tris–HCl, 300 mM NaCl, 5 mM $MgCl_2$, 0.5% (v/v) Triton X-100, 5 mM DTT (added prior to use), 1 tablet/50 ml protease inhibitor (added prior to use), make up to 50 ml with DEPC-treated water.

C. 2× SDS loading buffer (pH 6.8): 0.125 M Tris–HCl, 4% (w/v) SDS, 20% (v/v) glycerol, 2% (v/v) 2-mercaptoethanol, 0.05% (w/v) bromophenol blue, and distilled water.

D. Bromophenol blue	(ICN Biomedical)
E. Bradford reagent	(BioRad)
F. Protein A-Agarose	(Roche)
G. Anti-HA Affinity matrix	(clone 3 F10, Roche)
H. Glycogen blue	(Ambion)

2.3. RNA Extraction Using TRIzol Reagent

A. TRIzol reagent	(Invitrogen)
B. Chloroform	(Fisher)
C. 3 M sodium acetate (pH 5.2)	(Fisher)
D. 75 and 100% ethanol	(RNase free)
E. Glycogen blue	(Ambion)

2.4. Small RNA Library Construction

A. 0.3 M NaCl: In DEPC-treated water.

B. 3′ Linker: 5′-rAppCTGTAGGCACCATCAAT/3ddC/-3′.

C. 5′ Linker: 5′-TGGAATrUrCrUrCrGrGrGrCrArCrCrArArGrGrU-3′.

D. Agarose	(Fisher)
E. ATP	(NEB)
F. ATP-free T4 RNA ligase buffer	(NEB)
G. DMSO (RNase free)	(Fisher)
H. 5× First Strand buffer	(Invitrogen)

(continued)

(continued)

I. 10% Urea-PAGE: 2.5 ml 40% polyacrylamide (acrylamide:bis-acrylamide = 29:1), 1 ml 5× TBE (RNase free), 4.2 g urea (RNase free), 50 μl 10% ammonium persulfate (APS), 5 μl N,N,N',N'-tetramethylethylenediamine (TEMED), add water (DEPC treated) to 10 ml.	
J. dNTPs (10 mM; in DEPC-treated water)	(Fermentas)
K. DTT (0.1 M)	(Invitrogen)
L. Gel loading buffer II	(Ambion)
M. RNase-OUT	(Invitrogen)
N. SuperScript III reverse transcriptase	(Invitrogen)
O. Ethidium bromide (EtBr)	(Fisher)
P. GeneRuler 100 bp DNA ladder	(Fermentas)
Q. O'RangeRuler 10 bp DNA ladder	(Fermentas)
R. Polymerase chain reaction (PCR) 5′ primer: 5′-AATGATACGGCGACCACCGACAGGTTCA GAGTTCTAC AGTCCGA 3′.	
S. RT and PCR 3′ primer: 5′-CAAGCAGAAGACGGCATACGATTGA TGGTGCCTACAG-3′.	
T. PrimeSTAR HS DNA polymerase	(Takara)
U. T4 RLN2	(NEB)
V. T4 RNA ligase	(NEB)

2.5. mRNA Detection

A. DNase I	(Invitrogen)
B. 25 mM EDTA	(Invitrogen)
C. Oligo d(T) primer.	
D. *Taq* DNA polymerase	(NEB)
E. 6× DNA loading dye	(Fermentas)
F. 2% agarose gel.	

3. Methods

3.1. Materials and Reagents Prepared Prior to Experiments

A. Prepare IP extraction buffer and IP washing buffer without DTT and protease inhibitor cocktails. Prepare 50 mg/ml rifampicin and 50 mg/ml kanamycin stock solutions and sterilize by filtration.

B. Grow transgenic *3HA:GUS* (for AGO1-immunoprecipitation using anti-AGO1 antibody) and *AGO2::3HA:AGO2* (for AGO2-immunoprecipitation using HA antibody) *Arabidopsis* plants at a 12-h light/12-h dark photoperiod 4 weeks prior to starting experiments.

C. Add DTT and protease inhibitor cocktail to IP extraction buffer and IP washing buffer immediately before using.

D. Pre-equilibrate protein-A beads with IP extraction buffer immediately before using.

E. Pre-equilibrate conjugated HA beads with IP extraction buffer immediately before using.

3.2. Bacteria Pathogen Inoculation

A. Culture *Pst* (*avrRpt2*) on a PAF plate supplemented with antibiotics overnight at 28°C.

B. On the second day, collect the bacteria and resuspend in 10 mM $MgCl_2$ to a final concentration of $OD_{600}=0.02$ (see Note 1). Prepare 10 mM $MgCl_2$ solution as a mock control.

C. Syringe infiltrate the 4-week-old *3HA:GUS* and *AGO2::3HA:AGO2 Arabidopsis* plants with 10 mM $MgCl_2$ (mock) and *Pst* (*avrRpt2*). A needleless syringe is used to pressure infiltrate the abaxial (lower) side of the leaves with care such that plant tissue is not damaged.

D. Collect infiltrated leaves 12 h after inoculation and freeze immediately in liquid nitrogen. Afterward, tissue can be stored at −80°C for several months if not processed immediately.

3.3. Protein Immunoprecipitation (See Note 2)

A. Precool mortars and pestles with liquid nitrogen. Grind 0.25-g leaf tissue (see Note 3) into a fine powder in liquid nitrogen. Keep materials frozen all the time. Transfer ground tissue to a 1.5-ml microcentrifuge tube (precooled by liquid nitrogen prior to use) without thawing. Add 1.0 ml cold IP extraction buffer (see recipe in the Subheading 2) to the tube. Mix IP extraction buffer with ground plant tissue thoroughly by inverting and gentle shaking (no vortexing). Keep tubes on ice until no frozen tissue is visible.

B. Continue to mix by inverting for 10 min in a cold room.

C. Centrifuge the homogenate at $10,000 \times g$, 4°C, for 10 min. Pass the supernatant through a cell strainer (see Note 4).

D. Aliquot 10 μl supernatant or IP extraction buffer (background control) to a 1.5-ml microcentrifuge tube. Add 1 ml BradFord reagent to each tube. After mixing by inverting, leave at room temperature for 10 min, and then measure and record the optical density at 595 nm (OD_{595}). Adjust the volume of each sample by adding IP extraction buffer so that each sample will have equal OD units (OD_{595} × volume).

E. Save 10 μl as inputs for Western blotting detection later by adding 10 μl 2× SDS loading buffer (see recipe in the Subheading 2) and boiling for 5 min.

F. Preclear aliquots by adding 25 μl Protein A-Agarose beads to the protein extract corresponding to each 0.25-g plant tissue (pre-equilibrated). Rotate the tubes for 1 h in a cold room. Centrifuge at 200×g, 4°C, for 30 s to collect the beads and transfer the supernatant to new tubes.

G. Add 25 μl Anti-HA Affinity Matrix to the protein extract corresponding to each 0.25 g of plant tissue (pre-equilibrated). Rotate in a cold room for 2 h (see Note 5).

H. Collect beads by centrifuging at 200×g, 4°C, for 30 s.

I. Remove the supernatant. Save 10 μl as flow through by adding 10 μl 2× SDS loading buffer and boiling for 5 min.

J. Collect the Anti-HA Affinity Matrix beads into one tube if multiple tubes have been used. Add 1 ml IP washing buffer to each tube and mix by inverting five to ten times. Collect the beads by centrifuging at 200×g, 4°C, for 30 s and then removing the supernatant. Repeat once.

K. Add 1 ml IP washing buffer to each tube. Keep tubes in constant inverting motion for 10 min at 4°C. Collect the beads by centrifuging at 200×g, 4°C, for 30 s. Repeat this wash step three times (see Note 6).

L. Resuspend the beads in 1 ml IP washing buffer. Aliquot 200 μl to a new microcentrifuge tube. Collect beads by centrifuging at 200×g, 4°C, for 30 s. Add 50 μl 1× SDS loading buffer and boil for 5 min. This fraction will be used for quality control in SDS-PAGE and Western blotting analysis.

M. Collect the remaining beads by centrifuging at 200×g, 4°C, for 30 s. Completely remove the IP washing buffer over the beads by pipetting. The drained beads are now ready for further analysis (see Note 7).

3.4. RNA Extraction Using TRIzol Reagent (See Note 8)

A. Resuspend the beads in 300 μl cold IP washing buffer and keep the tubes on ice.

B. In a chemical hood, add 800 μl TRIzol reagent to each tube and mix the contents for 3 min by vortexing.

C. Add 300 μl chloroform to each tube and mix for 15 s by vortexing. Leave the tubes at room temperature for 3 min.

D. In a cold room, centrifuge the tubes at 15,000×g for 10 min.

E. Transfer the aqueous phase to new tubes without disturbing the organic and intermediate phases.

F. Add 15 mg glycogen blue, 1/10 volume of 3 M sodium acetate, and 2 volume of 100% ethanol (RNase free) to each tube. Mix the tubes by gently inverting five to six times.

G. Precipitate RNAs overnight at −20°C (see Note 9).

H. Spin the tubes at 15,000×g, 4°C, for 30 min. A tiny blue pellet should form at the tip of each tube.

I. Carefully remove the liquid, avoiding aspirating the pellets. Add 500 μl 70% ethanol (RNase free) to the tubes and wash the tubes by gently inverting five to six times.

J. Spin the tubes at 15,000×g, 4°C, for 5 min.

K. Completely remove the liquid, avoiding aspirating the pellets. Collect the remaining liquid to the bottom of the tubes by brief spinning. Remove the last drop until no liquid on the wall of the tubes is visible.

L. Air dry the pellet for 3–5 min (see Note 10).

M. Resuspend the pellet in 15 μl DEPC-treated water (see Note 11).

3.5. Small RNA Library Construction

3.5.1. 3′ Adaptation

A. On ice, set up a 3′ adaptation reaction (add in the following order):

RNA (from immunoprecipitation; in water)	13 μl
ATP-free T4 RNA ligase buffer (NEB; 10×)	2 μl
DMSO	2 μl
3′ adapter	1 μl
T4 RLN2 (NEB)	2 μl

B. Incubate the reaction at 25°C for 1 h.

C. Bring the volume to 100 μl with DEPC-treated water.

D. Follow steps F to L in Subheading 3.4 to precipitate RNA.

E. Resuspend the pellet in 20 μl DEPC-treated water.

3.5.2. Gel Purification of 3′ Adapted Small RNAs

A. Prepare a 10% polyacrylamide/urea gel as described in Subheading 2 (12).

B. Add equal volume of gel loading buffer II to each sample. Heat samples at 65°C for 5 min. Immediately chill in ice for 5 min.

C. Load the samples on the gel. In a separate well, load 10 μl 10 bp DNA ladder.

D. Run the gel for 2 h on constant voltage (15 V/cm) or until the bromophenol blue dye runs off the front of the gel.

E. In a clean (RNase free) container, stain the gel with DEPC-treated water and EtBr for 10 min on a platform rocker at room temperature.

F. Under UV light (see Note 12), carefully cut the gel between 30 and 50 bp (using the marker as a guide) and transfer the gel slices into a clean 1.5-ml microcentrifuge tube.

G. Using a sterile 1-ml tip, crush the gel slices into fine debris (see Note 13).

H. Add 400 μl 0.3 M NaCl to each tube.

I. Elute RNA from the gel by rotating overnight in a cold room.

J. Spin the tubes at $15,000 \times g$, 4°C, for 10 min to pellet the gel debris.

K. Carefully pipet the liquid into new tubes without carrying over any gel debris (see Note 14).

L. Follow steps F to L in Subheading 3.4 to precipitate RNA.

M. Resuspend the pellet in 11 μl DEPC-treated water.

3.5.3. 5′ Adaptation

A. On ice, set up a 5′ adaptation reaction (add in the following order):

3′-ligated RNA (from previous step, in water)	11 μl
T4 RNA ligase buffer (10×)	2 μl
ATP (10 mM)	2 μl
DMSO	2 μl
5′ adapter	1 μl
T4 RNA ligase (NEB)	2 μl

B. Incubate the reaction at 37°C for 1 h.

C. Bring the volume to 100 μl by adding DEPC-treated water.

D. Follow steps F to L in Subheading 3.4 to precipitate RNA.

E. Resuspend the pellet in 20 μl DEPC-treated water.

3.5.4. Gel Purification of 3′- and 5′-Adapted Small RNAs

A. Prepare a 10% polyacrylamide/urea gel as described in Subheading 2 (12).

B. Add equal volume of gel loading buffer II to each sample. Heat samples at 65°C for 5 min. Immediately chill in ice for 5 min.

C. Load the samples on the gel. In a separate well, load 10 μl 10 bp DNA ladder.

D. Run the gel for 2 h at constant voltage (15 V/cm) or until the bromophenol blue dye runs off the front of the gel.

E. In a clean (RNase free) container, stain the gel with DEPC-treated water and EtBr for 10 min on a platform rocker at room temperature.

F. Under UV light (see Note 12), carefully cut the gel between 50 and 100 bp (using the marker as a guide) and transfer the gel slices into a clean 1.5-ml microcentrifuge tube.

G. Using a sterile 1-ml tip, crush the gel slices into fine debris (see Note 13).

H. Add 400 μl 0.3 M NaCl to each tube.

I. Elute RNA from the gel by rotating overnight in a cold room.

J. Spin the tubes at 15,000×g, 4°C, for 10 min to pellet the gel debris.

K. Carefully pipet the liquid into new tubes without carrying over any gel debris (see Note 14).

L. Follow steps F to L in Subheading 3.4 to precipitate RNA.

M. Resuspend the pellet in 20 μl DEPC-treated water.

3.5.5. Reverse Transcription of 3′- and 5′-Adapted Small RNAs

A. On ice, set up a reverse transcription reaction (add in the following order):

3′- and 5′-ligated RNA (gel purified; in water)	11 μl
Reverse transcription primer (SBS3; 10 μM)	1 μl
dNTPs (10 mM)	1 μl

B. Incubate at 65°C for 5 min.

C. Place on ice and add:

5× first-strand buffer	4 μl
0.1 M DTT	1 μl
RNase-OUT (40 U/μl)	1 μl
SuperScript III RT (200 U/μl)	1 μl

D. Incubate at 50°C for 1 h and then at 70°C for 15 min (see Note 9).

3.5.6. PCR Amplification of Small RNA Library

A. On ice, set up a PCR reaction (add in the following order):

PrimeSTAR buffer (5×)	10 μl
dNTPs (2.5 mM)	4 μl
PCR 5′ primer (10 μM)	1 μl
PCR 3′ primer (10 μM)	1 μl
First-strand cDNA (from reverse transcription)	1–5 μl
PrimeSTAR HS DNA polymerase (2.5 units/μl)	0.5 μl
Sterile water	To 50 μl

B. Set the following program on a thermocycler:

| 98°C for 10 s |
| 55°C for 5 s |
| 72°C for 10 s |
| Amplify for 22 cycles |

C. Load samples onto a 10% polyacrylamide gel (0.5× TBE without urea) and separate the samples by running the gel for 2 h at constant voltage (15 V/cm).

D. In a clean container, stain the gel with distilled water and EtBr for 10 min on a gentle rocker at room temperature.

E. Under UV light (see Note 12), carefully cut the gel at 110 bp (using the marker as a guide) and transfer the gel slices into a clean 1.5-ml microcentrifuge tube (see Note 15).

F. Using a sterilized 1-ml tip, crush the gel slices into fine debris.

G. Add 400 μl 0.3 M NaCl to each tube.

H. Elute RNA from the gel by rotating overnight in a cold room.

I. Spin the tubes at $15,000 \times g$, 4°C, for 10 min to pellet the gel debris.

J. Carefully pipet the liquid into new tubes without carrying over any gel debris (see Note 14).

K. Follow steps F to L in Subheading 3.4 to precipitate RNA (see Note 16).

L. Resuspend the pellet in 10 μl DEPC-treated water.

4. Notes

1. A bacteria optical density of 0.02 (at 600 nm) is equivalent to 2×10^7 colony forming units per milliliter (cfus).

2. All procedures should be carried out on ice or in a cold room unless indicated otherwise. Cool tubes and tips in a cold room prior to use.

3. Begin with 0.1–0.2-g plant tissue for a standard Northern blot. For constructing a small RNA library, 1–2-g tissue is needed.

4. Any gel debris carried over to the next step is going to affect efficient RNA precipitation and resuspension. In order to prevent carrying over gel debris to the new tubes, always leave a little bit of liquid on the top of the pellet. Repeating the spinning step one time is also useful to eliminate unwanted carryover.

5. This step could be extended overnight in a cold room if maximum pull-down efficiency is necessary. This should be done with caution if heavy RNase activity is observed in the extraction.

6. Washing may be repeated more times if nonspecific proteins remain in the immunoprecipitate.

7. The beads can be boiled with SDS-loading buffers for Western detection, or RNA can be extracted using TRIzol reagent or other similar reagents. This protocol should yield sufficient RNA for constructing a standard small RNA library for Illumina deep sequencing (Subheadings 3 and 4). The resulting RNA can also be used for Northern blotting or for investigating the target protein-associated RNAs.

8. For extracting RNAs associated with specific target proteins, all the tips and tubes should be treated with DEPC prior to use; all the reagents should be RNase free.

9. Experiment can be stopped at this step by storing reactions in a −80°C freezer.

10. Avoid overdrying that may cause difficulty in resuspending the sample.

11. 1–2 μl sample can be used for quality control from this step.

12. Use longer wavelength when possible (e.g., 365 nm instead of 302 nm). Exposure to high-energy UV light may cause damage to RNA samples and could potentially jeopardize cloning efficiency.

13. Always use autoclaved RNase-free tips. Contamination introduced from this step may cause RNA degradation during the following overnight incubation.

14. Any gel debris carried over to the next step is going to affect efficient RNA precipitation and resuspension. In order to prevent carrying over gel debris to the new tubes, always leave a little bit of liquid on the top of the pellet. Repeating the spinning step one time is also useful to eliminate unwanted carryover.

15. Care should be taken at this step so that only the cloned small RNAs, and not the self-ligation products, are collected. These two molecules have only a 20-nucleotides difference in length. Contamination at this step will affect downstream sequencing quality.

16. The pellet in this step may be invisible. Care should be taken at the washing step to avoid losing the pellet.

Acknowledgments

We thank Jim Carrington for providing the *HA:GUS* and *pAGO2:HA-AGO2* lines. The research work is supported by an NIH R01 GM093008, an NSF Career Award MCB-0642843, and an AES-CE Research Allocation Award PPA-7517H to H. Jin.

References

1. Ghildiyal M, Zamore PD (2009) Small silencing RNAs: an expanding universe. Nat Rev Genet 10(2):94–108
2. Mallory AC, Vaucheret H (2006) Functions of microRNAs and related small RNAs in plants. Nat Genet 38:S31–S36
3. Inui M, Martello G, Piccolo S (2010) MicroRNA control of signal transduction. Nat Rev Mol Cell Biol 11(4):252–263
4. van Rooij E et al (2007) Control of stress-dependent cardiac growth and gene expression by a MicroRNA. Science 316(5824):575–579
5. Voinnet O (2009) Origin, biogenesis, and activity of plant MicroRNAs. Cell 136(4):669–687
6. Vaucheret H (2008) Plant ARGONAUTES. Trends Plant Sci 13(7):350–358
7. Zhang XM et al (2011) Arabidopsis argonaute 2 regulates innate immunity via miRNA393*-mediated silencing of a golgi-localized SNARE gene, MEMB12. Mol Cell 42(3):356–366
8. Harvey JJ et al (2011) An antiviral defense role of AGO2 in plants. PLoS One 6(1):e14639
9. Mi S et al (2008) Sorting of small RNAs into Arabidopsis argonaute complexes is directed by the 52 terminal nucleotide. Cell 133(1):116–127
10. Montgomery TA et al (2008) Specificity of ARGONAUTE7-miR390 interaction and dual functionality in TAS3 trans-acting siRNA formation. Cell 133(1):128–141
11. Katiyar-Agarwal S, Jin HL (2007) Discovery of pathogen-regulated small RNAs in plants. In: Rossi JJ, Hannon GJ (eds) Academic Press; San Diego, California. Methods Enzymol 427:215–227
12. Sambrook, J. (2001) Molecular cloning: a laboratory manual Cold Spring Harbor, N.Y.: Cold Spring Harbor Laboratory Press

Chapter 14

New Virus Discovery by Deep Sequencing of Small RNAs

Kashmir Singh, Ravneet Kaur, and Wenping Qiu

Abstract

Small RNAs (sRNAs) have emerged as one of the most important regulators of gene expression in eukaryotes. sRNAs are intermediate molecules as well as end products in the antiviral defense pathway called RNA interference in plants and animals. Profiling of sRNAs using next-generation sequencing technologies has identified a number of plant viruses that have never been reported previously, and has provided a deeper view of virus populations in a plant that cannot be achieved by conventional methods like PCR and ELISA. In this chapter, we describe the methodology of deep sequencing of sRNAs. The high-throughput and highly sensitive method will revolutionize the identification of plant viruses and the study of molecular plant–virus interactions.

Key words: RNAi, Small RNAs, Deep sequencing, Virus identification, Bioinformatics

1. Introduction

RNA interference (RNAi) is the process of sequence-specific degradation of target RNA molecules triggered by double-stranded RNA (dsRNA) (1). Depending on the organism and the source of the dsRNA trigger, RNAi normally results in transcriptional and/or posttranscriptional gene silencing. Posttranscriptional RNA silencing is achieved by either mRNA degradation or translational repression. Some of the functions of RNAi processes are well known. For example, in fission yeast, RNAi is one of the systems that establishes and maintains the heterochromatin structure of the centromere and mating-type locus (2). RNAi is also an antiviral process to keep virus infection under check in an organism (3). Small RNAs (sRNAs) comprise the sequence-specific effectors of RNAi pathways that direct the negative regulation or control of genes, repetitive sequences, mobile elements, and viruses (4, 5).

sRNAs are now recognized as a large class of gene regulators expressed in all eukaryotes, including unicellular organisms such as *Chlamydomonas reinhardtii* (1, 6, 7). In plants, these small regulatory RNAs consist of microRNAs (miRNAs) and small interfering RNAs (siRNAs), which can be differentiated by their distinct modes of biogenesis and the types of genomic loci from which they are derived. miRNAs are short, endogenously expressed, non-translated RNAs that are processed by Dicer-like proteins from stem-loop regions of longer pre-miRNA precursors. Like miRNAs, siRNAs are processed by the Dicer RNase III family of enzymes, but instead of deriving from local stem-loop structures, siRNAs are processed from long, double-stranded RNA precursors (either from much longer stems or from bimolecular duplexes of a viral genome). Although miRNAs and siRNAs are derived from different types of molecules, they are chemically similar. Both miRNAs and siRNAs are incorporated into the silencing complexes that contain ARGONAUTE proteins, wherein they can guide repression of target genes.

Many endogenous siRNAs are generated as a result of virus infection and play important roles in reprogramming of gene expression in plant defense responses. To gain insights into the total population and better understand siRNA's function in plants, a number of researchers turned to sequencing the sRNA component of the plant transcriptome (smRNAome). Initial discovery of miRNAs in the model plant Arabidopsis and rice used only small-scale sequencing of sRNA libraries (8–10). Now, the rapid "next-generation" DNA sequencing technologies at reasonably low cost have dramatically advanced our capacity to comprehensively understand the nucleic acid-based information in a cell at unparalleled resolution and depth. This technology has been employed to study genome sequence variation, ancient DNA, DNA methylation, protein–DNA interactions, transcriptomes, alternative splicing, mRNA regulation, and virus populations. Current deep sequencing technologies produce many gigabases of sequence at single-base resolution and can perform multiple genome-scale experiments in a single run, thus being effective in the analysis of many plant genomes. The deep sequencing of RNA has accelerated the identification of numerous novel sRNAs expressed at low abundance in several plant species (11, 12).

The deep sequencing of RNA technology is widely utilized to identify new viruses in infected plants by analyzing siRNA populations. RNAi is a cytoplasmic surveillance system for degradation of single-stranded RNA (ssRNA) and double-stranded RNA molecules homologous to the inducer (a virus in this case) by using siRNA as a guide (13). Virus infection triggers RNAi, and then becomes a target of RNAi (13). As a result of RNAi, viral siRNAs accumulate abundantly in infected cells. The progressive degradation of the invading viral genome at several locations and the biogenesis of siRNAs result in the generation of different types of siRNAs with various lengths of 21–24 nucleotides (14). The deep sequencing of siRNAs has led to identification of novel viruses (15–17) as well as commonly known viruses in symptomless plants that have extremely low virus titer (18). This unbiased sequencing strategy indeed has revealed the diversity, population structure, and genetic composition of viral genomes in a single plant.

In this chapter, we describe the major experimental approaches including high-throughput parallel sequencing and further analysis for the discovery and detection of virus-derived sRNAs. We applied the deep sequencing of sRNA technology to identify viruses that are associated with the vein-clearing and vine decline syndrome in a grapevine. We discovered *Grapevine vein clearing virus* (*GVCV*) and other viruses that have never been reported previously in grapevines (19).

2. Materials

Prepare all solutions using ultrapure water (prepared by purifying deionized water to attain a sensitivity of 18 MΩ at 25°C) and analytical grade reagents.

2.1. RNA Isolation

1. For RNA isolation, water is always treated with 0.1% (v/v) diethyl pyrocarbonate (DEPC) following standard procedure as detailed by Sambrook et al. (20). All the glassware and plasticware should be free from RNases (see Note 1).
2. Solution I: Phenol saturated with Tris (hydroxymethyl) aminomethane buffer to a pH of 6.7±0.2, sodium dodecyl sulfate [SDS; 0.1% (w/v)], sodium acetate [NaOAc; 0.32 M], and ethylenediamine tetra acetic acid (EDTA; 0.01 M final concentration from a 0.5 M stock solution, pH 8.0) (see Note 2).
3. Chloroform.
4. 70% Ethanol.
5. Isopropanol.
6. DNase I.
7. Phenol:chloroform (24:1).
8. 3 M NaOAc, pH 4.8.
9. Agarose.
10. 5× FA buffer: Dissolve 20.6 g of MOPS in 800 ml of DEPC-treated 50 mM sodium acetate. Adjust to pH 7.0 with 2 N NaOH and add 10 ml of DEPC-treated EDTA (0.5 M; pH 8.0). Adjust final volume to 1 L with RNase-free water.
11. Formaldehyde.

2.2. Size Fractionation of RNA and Elution of Small RNAs

1. TBE: 54 g Trizma base, 27.5 g H_3BO_3, 20 ml 0.5 M EDTA, pH 8.0.
2. 2× RNA loading buffer: 96% Formamide, 18 mM EDTA, 0.025% SDS, 0.025% saturated bromophenol blue (BPB), 0.025% saturated xylene cyanole.
3. 15% PAGE gel: 24 g urea, 5 ml 10×TBE, 18.75 ml 40% acrylamide/bisacrylamide solution 19:1, H_2O to 50 ml. Filter it and add 0.5 ml 10% APS and 16 µl TEMED for polymerization (see Note 3).
4. 10 bp ladder.
5. Ethidium bromide.
6. 0.4 M NaCl.
7. Absolute alcohol.

2.3. Library Preparation

1. Small RNA sample prep Kit, Illumina, USA.
2. T4 RNA Ligase 2, truncated, with 10× T4 RNL2 truncated reaction buffer.
3. 10 mM ATP.
4. 100 mM $MgCl_2$.
5. Ethanol (100, 75, and 70%).
6. Clean scalpels.
7. 21-gauge needles.
8. SuperScript II reverse transcriptase with 100 mM DTT and 5× first-strand buffer.
9. 6% TBE PAGE gel.
10. 3 M NaOAc, pH 5.2.
11. 6× DNA loading dye.
12. Nuclease-free water.

2.4. Validation of Library and Solexa Sequencing

1. Agilent Bioanalyzer 2100.
2. *Taq* polymerase with compatible buffer and $MgCl_2$.
3. Illumina GAIIx sequencing platform (service mostly utilized from commercial service provider).

2.5. Bioinformatics Analysis

1. Computer with suitable operating system.
2. Software or available online tools to remove 3′ and 5′ adapters.
3. Program to assemble short sRNA reads into longer fragments (contigs).
4. BLAST program.
5. Database of viruses and viroids.

2.6. PCR Analysis to Confirm the Presence of a Virus

1. Software to align reads on a reference genome.
2. Primers specific to viruses identified by BLAST analysis.
3. *Taq* polymerase.
4. T/A cloning vector.
5. Competent cells for transformation.

3. Methods

3.1. RNA Isolation

This protocol utilizes the phenol-based extraction method, which allows the rapid isolation of RNA (21). The conditions enable the partitioning of proteins and DNA into the organic layer of the biphasic solution interface while retaining RNA in the upper aqueous layer. The aqueous phase is removed to a second tube, and RNA is precipitated with an equal volume of isopropanol. High yields of pure, undegraded total RNA can be recovered from small quantities of tissue or cells. Large numbers of samples can be processed simultaneously because of the simplicity of the technique.

1. Grind 100 mg of tissue to a fine powder in liquid nitrogen using a mortar and pestle.
2. Add 2 ml of solution I. Solution I gets frozen as added; continue grinding so as to make a homogenous mixture; this ensures close contact of the tissue with the reagents that would help instantaneous denaturation of protein. Allow the mixture to thaw completely with intermittent grinding.
3. Add 800 μl of autoclaved DEPC-treated water, and mix it by grinding.
4. Transfer the contents to two 2-ml microcentrifuge tubes and leave them at room temperature for 5 min.
5. Add 200 μl of chloroform to each tube, vortex briefly (<10 s), and leave for 10 min at room temperature.
6. Centrifuge at $11,000 \times g$ for 10 min at 4°C and transfer the upper aqueous phase into fresh tubes.
7. Add 0.6 vol. of isopropanol, vortex briefly (<10 s), and leave for 10 min at room temperature (see Note 4).
8. Centrifuge at $11,000 \times g$ for 10 min at 4°C and discard the supernatant.
9. Wash RNA pellet with 70% ethanol, air dry, and dissolve in 20–50 μl of DEPC-treated water (see Note 5).
10. Add 1 unit of DNaseI for 1–5 μg of RNA and incubate for 30 min at 37°C (see Note 6).
11. Purify the RNA first by extraction with phenol:chloroform (24:1), then with chloroform, and precipitate by adding 3 M

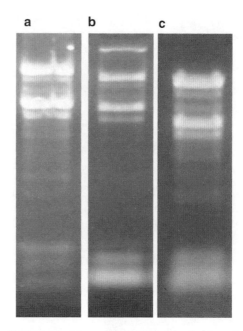

Fig. 1. Isolation of RNA from leaf tissue. (**a**) RNA isolated without small RNA enrichment; (**b**) RNA enriched for small RNAs but with DNA contamination; (**c**) clean RNA enriched for small RNAs after digestion with DNase I. Following the enrichment protocol, an intense small RNA band can be seen in (**b**) and (**c**) as compared to (**a**).

sodium acetate (pH 4.8) and chilled ethanol. Centrifuge at $11,000 \times g$ for 10 min at 4°C and discard the supernatant.

12. Wash the pellet with 70% ethanol, air dry, and dissolve in RNase-free water.

13. Check the purity and concentration of RNA by determining the absorbance of the sample at 260 and 280 nm using a spectrophotometer.

14. Check the integrity of RNA by running on a denaturing formaldehyde-agarose gel (20) using FA buffer (Fig. 1).

3.2. Size Fractionation of RNA and Elution of Small RNAs

1. Clean glass plates, combs, spacers, and tank by washing them in 10% SDS, and then sterile water. Additionally, wash plates with 95% ethanol.

2. Prepare 15% PAGE gel. Mix well each time, and pour the uniform solution between prepared glass plates. Insert comb and place plates almost horizontally. The gel should be polymerized in 20 min.

3. Fasten plates to electrophoresis tank and fill it with 1× TBE buffer. Wash the wells thoroughly with TBE to remove urea. Run for 30 min at 200–250 V before loading the samples. During this time, prepare RNA samples.

4. Mix RNA (1–10 µg) with an equal volume of RNA loading buffer and denature at 70°C for 10 min. Wash the wells and

load your samples with a thin pipette tip carefully onto the flat bottom of the well (see Note 7). Run the gel at 200–250 V. As a marker, use 10 bp ladder (Decade marker). Label the marker with radiolabeled ^{32}P before loading using T4 polynucleotide kinase or any other labeling procedure.

5. Stop the electrophoresis when BPB (leading dye) is about 1.5 cm above the bottom of the gel to see the 10-mer band (see Note 8).

6. For isolation of small RNAs of desired size (22), dismantle the gel, leaving it mounted on one glass plate. To facilitate the alignment of the gel to the phosphorimager paper printout, implant tiny fluorescent paper pieces asymmetrically at three of the four corners of the gel. Wrap the gel in plastic film (e.g., Saran wrap) and expose it to a phosphorimaging screen for 45 min. Print out a 100%-scaled image of the gel, and align the gel on top of the printout according to the position of the three radioactive gel pieces.

7. Excise the bands defined by the mobility of the RNA size markers. Cut the gel slice in smaller pieces so that they can fit into a preweighed siliconized 1.5-ml tube. Determine the weight of the gel slices. Add 2–3 vol. (v/w) of RNase-free 0.4 M NaCl and elute the small RNAs from the gel by incubating the tube overnight at 4°C under constant agitation. Collect the supernatant and precipitate the small RNAs for at least 2 h on ice or overnight at −20°C after the addition of 3–4 vol. of absolute ethanol.

8. Collect the small RNA pellet after ethanol precipitation in a tabletop centrifuge for 15 min at maximum speed ($\sim$14,000 $\times g$) at 4°C. Remove the supernatant and collect the residual liquid at the bottom of the tube by an additional 5-s centrifuge spin. Remove the residual liquid completely using a small pipette tip without perturbing the pellet. The additional spin is needed to collect all residual liquid. If necessary, air dry the RNA pellet to evaporate residual ethanol. Be sure that the ethanol has been evaporated as it may inhibit the subsequent enzymatic steps. Dissolve the RNA pellet in RNase-free water.

3.3. Library Preparation

For library preparation, we used a kit supplied by Illumina (Small RNA Sample Prep Kit FC-102-1009). The small RNA 3′ adapter in the kit is specifically modified to target miRNAs and other sRNAs that have a 3′ hydroxyl group resulting from enzymatic cleavage by Dicer or other RNA-processing enzymes. The 3′ adapter is required for reverse transcription and corresponds to the surface-bound amplification primer on the flow cell. The 5′ small RNA adapter is necessary for amplification of the small RNA fragments. Primers are also needed to perform reverse transcription and PCR (Table 1).

Table 1
Sequences of adapters and various primer sequences used for library preparation and sequencing

Primer and adapters	Sequence of primers and adapters
RT primer	5′ CAAGCAGAAGACGGCATACGA
5′ RNA adapter	5′ GUUCAGAGUUCUACAGUCCGACGAUC
3′ RNA adapter	5′ P-UCGUAUGCCGUCUUCUGCUUGUidT
Small RNA PCR primer 1	5′ CAAGCAGAAGACGGCATACGA
Small RNA PCR primer 2	5′ AATGATACGGCGACCACCGACAGGTTCAGAGTTCTACAGTCCGA

3.3.1. Ligation of Small RNA 3′ Adapter and 5′ Adapter

1. Set up the ligation reactions in a sterile, nuclease-free, 200-μl microcentrifuge tube using the following:
   Total RNA in nuclease-free water (5.0 μl)
   Diluted 1× sRNA 3′ adapter (1 μl)
2. Incubate at 70°C for 2 min, and then transfer immediately to ice.
3. Add the following reagents and mix well:
   10× T4 RNL2-truncated reaction buffer (1 μl)
   100 mM $MgCl_2$ (0.8 μl)
   T4 RNA Ligase 2 truncated (NEB) (1.5 μl)
   RNaseOUT (0.5 μl)
4. Incubate at 22°C for 1 h.
5. With 5 min remaining, prepare the 5′ adapter for ligation by heating it at 70°C for 2 min, and then transferring it to ice.
6. Add the following reagents to the ligation mixture from step 4 and mix well:
   10 mM ATP (1 μl).
   sRNA 5′ adapter (0.5 μl).
   T4 RNA ligase (1 μl).
7. Incubate at 20°C for 1 h. Store at 4°C.

3.3.2. Reverse Transcription and Amplification of sRNA Ligated with Adapters

Reverse transcription followed by PCR is used to create cDNA constructs of the small RNA ligated with 3′ and 5′ adapters. This selectively enriches those fragments that have adapter molecules on both ends. The PCR is performed with two primers that anneal to the ends of the adapters.

1. Combine the following in a sterile, nuclease-free, 200-μl microcentrifuge tube:

PAGE-purified 5′ and 3′ ligated RNA (4.0 μl).

Diluted sRNA RT primer (1.0 μl).

The total volume should be 5 μl.

2. Heat the mixture at 65°C in a thermal cycler for 10 min.
3. Place the tube on ice.
4. Premix the following reagents:

   5× First-strand buffer (2 μl).

   mM dNTP mix (0.5 μl).

   100 mM DTT (1 μl).

   RNaseOUT (0.5 μl).

5. Add 4 μl of the mix to the cooled tube containing the primer-annealed template material. The total volume should now be 9 μl (5 μl of template preparation and 4 μl of reverse transcription).
6. Heat the sample to 48°C in a thermal cycler for 3 min.
7. Add 1 μl SuperScript II Reverse Transcriptase. The total volume should now be 10 μl.
8. Incubate in a thermal cycler at 44°C for 1 h.
9. Prepare the PCR Master Mix:

   Ultrapure water (27 μl).

   5× Phusion HF buffer (10 μl).

   Small RNA primer 1 GX1 (1.0 μl).

   Small RNA primer 2 (1.0 μl).

   25 mM dNTP mix (0.5 μl).

   Phusion DNA polymerase (0.5 μl).

   The total volume should be 40 μl.

10. Perform PCR amplification.

    Add 40 μl of PCR master mix into a sterile, nuclease-free, 200-μl PCR tube.

    Add 10 μl of single-strand reverse-transcribed cDNA.

    Amplify the PCR in the thermal cycler using the following PCR protocol: 30 s at 98°C followed by 9–12 cycles of 10 s at 98°C, 30 s at 60°C, and 15 s at 72°C, then hold for 10 min at 72°C, and finally hold at 4°C.

3.3.3. Purification of Amplified cDNA Products and Selection of sRNA Libraries

This protocol gel purifies the amplified cDNA construct in preparation for loading on the Illumina Cluster Station. Assemble the gel electrophoresis apparatus as per the manufacturer's instructions.

1. Mix 1 μl of 25 bp ladder with 1 μl of 6× DNA loading dye.
2. Mix 50 μl of amplified cDNA construct with 10 μl of 6× DNA loading dye.

3. Load 2 µl of mixed 25 bp ladder and loading dye in one well on the 6% PAGE gel.
4. Load two wells with 25 µl each of mixed amplified cDNA construct and loading dye on the 6% PAGE gel.
5. Run the gel for 30–35 min at 200 V or until the front dye exits the gel.
6. Remove the gel from the apparatus.
7. Puncture the bottom of a sterile, nuclease-free, 0.5-ml microcentrifuge tube four to five times with a 21-gauge needle.
8. Place the 0.5-ml microcentrifuge tube into a sterile, round-bottom, nuclease-free, 2-ml microcentrifuge tube.
9. Pry apart the cassette and stain the gel with the ethidium bromide in a clean container for 2–3 min.
10. View the gel on a Dark Reader transilluminator or a UV transilluminator.
11. Using a clean scalpel, cut out the bands corresponding to approximately the adapter-ligated constructs derived from the 22- and 30-nt small RNA fragments. The band containing the 22-nt RNA fragment with both adapters will be a total of 93 nt in length. The band containing the 30-nt RNA fragment with both adapters will be 100 nt in length.
12. Place the band of interest into the 0.5-ml microcentrifuge tube from step 7.
13. Centrifuge the stacked tubes at 14,000 rpm in a microcentrifuge for 2 min at room temperature to move the gel through the holes into the 2-ml tube.
14. Add 100 µl of 1× gel elution buffer to the gel debris in the 2-ml tube.
15. Elute the DNA by rotating the tube gently at room temperature for 2 h.
16. Transfer the eluate and the gel debris to the top of a Spin-X filter.
17. Centrifuge the filter for 2 min at 14,000 rpm.
18. Add 1 µl of glycogen, 10 µl of 3 M NaOAc, and 325 µl of −20°C 100% ethanol.
19. Immediately centrifuge to 14,000 rpm for 20 min in a benchtop microcentrifuge.
20. Remove and discard the supernatant, leaving the pellet intact.
21. Wash the pellet with 500 µl of room-temperature 70% ethanol.
22. Remove and discard the supernatant, leaving the pellet intact.
23. Dry the pellet using the speed vac.
24. Resuspend the pellet in 10 µl resuspension buffer.

3.4. Validation of Library and Solexa Sequencing

It is recommended to perform quality control analysis of sample library by loading 1 µl of the resuspended construct on an Agilent Technologies 2100 Bioanalyzer and check the size, purity, and concentration of the sample. Quality-passed samples can be used for sequencing. For small RNAs, generally, 36-cycled single-end sequencing is preferred.

Solexa sequencing: Perform the following PCRs with lower primer concentrations (100 nM) to eliminate the need for removal of unincorporated primer oligodeoxynucleotides.

1. Prepare a 1:100 dilution of the DNA obtained in step 24 above.
2. Use 10 µl of the dilution for a 100 µl pilot standard PCR using *Taq* polymerase (100 bnM primers, 2 mM $MgCl_2$) with the following cycle conditions: 45 s at 94°C, 1 min 25 s at 50°C, 1 min at 72°C. Take aliquots after 6, 8, 10, 12, and 14 cycles and analyze them on a 2% agarose gel.
3. Select the number of cycles before reaching saturation, typically after eight cycles.
4. Perform a 100 µl (10 µl sample, 100 nM primers) PCR with optimal number of cycles and analyze it on an agarose gel.
5. The sample is now ready for entering the Solexa's sequencing step. Follow the manufacturer's instruction wherever needed. Typically, 36-cycle single-end sequencing is sufficient for analysis of sRNA populations.

3.5. Bioinformatics Analysis

Ten to thirty million bases of sequences are normally generated by Illumina sequencing. sRNA sequences generated by this procedure contain 3′ and sometimes 5′ adapter sequences. The adapter sequences must be removed before further analysis. After adapter removal, short fragments are assembled into larger contigs. These contigs are searched against viral sequences present in the database using sequence similarity search programs like Basic Local Alignment Search Tool (BLAST) (23). This section discusses some of the software available to perform these analyses.

3.5.1. Adapter Removal

Adapter sequences attached to small cDNA sequences can be removed by using Novoindex and Novoalign software (Novocraft Technologies, Selangor Darul Ehsan, Malaysia). The programs run on a 64-bit computer with a minimum of 8 GB Random access memory (RAM) and Linux as operating system. An indexed reference genome is built first with Novoindex and then sRNA reads are aligned using Novoalign software (see Note 9).

Another program available online is UEA sRNA Toolkit and can be accessed freely using the Web interface http://srna-tools.cmp.uea.ac.uk/. This is an excellent online tool for adapter removal and miRNA analysis.

3.5.2. Contig Assembly

For assembling the short reads into longer contigs, SSAKE (24) and Velvet (25) are two reliable programs. Both programs can be run on user-defined parameters. The minimum computer specification for both programs is the same as that for Novocraft (see Subheading 3.5.1, step 1). SSAKE can assemble a large number of contigs with short sequences (see Note 10), whereas Velvet can assemble sRNA up to 3 kb in length. The sequences that are not assembled into contigs remain as singletons and can also be used in further analysis. Another advantage of assembling the short reads is to reduce the redundancy. After generating contigs, the next step is to search for sequence similarity using BLAST program (see Note 11).

3.5.3. BLAST Analysis

1. Install a stand-alone BLAST program available through the NCBI Web site (http://www.ncbi.nlm.nih.gov) on a laboratory computer (Linux operation system works better than Windows system).

2. Download the genome sequences of viruses and viroids that are available in the NCBI database using its FTP site.

3. Follow instructions as per the BLAST manual for database construction and use the scripts of BLAST. Upload the assembled contigs and singlet sequences to search the virus and viroid databases by using BLASTN and BLASTX.

4. The E-value cutoff should be kept low (the closer to zero, the better) to avoid mismatches and false predictions. BLAST generally can predict the type of viruses present in a sample.

3.6. PCR Analysis to Confirm the Presence of a Virus

The Mapping and Assembly with Qualities program (MAQ, http://maq.sourceforge.net) (15) can be used to determine the coverage and distribution of virus-specific small RNAs. This program aligns the small RNA reads (not contigs) to the genome of a reference virus. Regions of sequence alignments can then be used to design primers to amplify the viral fragments by PCR.

1. Download the reference genome of a virus, which showed high hits in BLAST, from the NCBI database and use it to align sRNA reads using the MAQ pipeline scripts.

2. Run MAQ using the easyrun script with default parameters. It accurately aligns the reads to various regions of the genome and gives information about the number of reads aligned to various locations. This alignment helps in identifying hot spots of the plant RNAi response against the virus.

3. Design forward and reverse PCR primers based on hot spots information, prepare the PCR mix, and run the PCR using thermal cycles.

4. Run the PCR product on 1.2% agarose gel. Visualize the amplified bands under UV and elute the bands from the gel.

5. Ligate the eluted PCR product into a T/A cloning vector and transform *E. coli*-competent cells. Isolate the plasmids and sequence them. Sequence analysis by BLAST will further confirm the presence of the virus in host tissue.

6. Do the primer walking to amplify and sequence the whole genome of a virus.

Using the method described above, we were able to identify 27 viruses from a single grapevine. Some of them were already reported from grapevine. However, we were able to identify a new DNA virus belonging to the badnavirus family (19). The entire genome of the virus was sequenced and submitted to GenBank.

4. Notes

1. The process needs very-high-quality RNA as starting material, with RNA integrity number (RIN) ranging from 8 to 10. Thus, all the glassware and plasticwares should be free from RNases, and solutions should be prepared in RNase-free water.

2. Proper care should be taken while weighing SDS as it is a strong denaturant. SDS precipitates at 4°C. Therefore, the lysis buffer needs to be warmed prior to use.

3. Wear a mask when weighing acrylamide. To avoid exposing coworkers to acrylamide, cover the weigh boat containing the weighed acrylamide with another weigh boat (similar size to the original weigh boat containing the weighed acrylamide) when transporting it to the fume hood. Transfer the weighed acrylamide to a cylinder inside the fume hood and mix on a stir plate placed inside the hood. Unpolymerized acrylamide is a neurotoxin and care should be exercised to avoid skin contact. Mixed bed resin AG 501-X8 (anion and cation exchange resin) is used when acrylamide solution is made, since it removes charged ions (e.g., free radicals) and allows longer storage. Filter out the resin before storage. The used mixed resin should be disposed as hazardous waste. Manufacturer's warning states that this resin is explosive when mixed with oxidizing substances. The resin contains a dye that changes from bluegreen to gold when the exchange capacity is exhausted.

4. The RNA isolation procedure described above generally provides very-high-quality RNA. However, in some tissues with extremely low sRNA populations, the enrichment of small RNAs can be done by adding 3 μl of glycogen (10 mg/ml) during precipitation of RNA.

5. Total RNA isolation can also be performed using the standard Tri-reagent (Sigma, USA), Trizol protocol (Invitrogen, USA), or the newly developed kits for sRNA extraction (also at higher prices per sample). *mir*Vana™ miRNA Isolation Kit (Ambion, USA) and miRNeasy Mini Kit (Qiagen, Germany) are two commercially available kits to isolate enriched high-quality small RNAs. miRNeasy Mini Kit provides an additional advantage because an option for on-column DNA digestion with DNase I is available that prevents the loss of RNA during further DNA digestion steps. Phenol-based isolation procedures can recover RNA species in the 10- to 200-nucleotide range (e.g., siRNAs, miRNAs, 5 S rRNA, 5.8 S rRNA, and U1 snRNA). Extraction procedures like Trizol/TriReagent, however, will purify all RNAs, large and small, and are recommended for isolating total RNA from biological samples that will contain miRNAs/siRNAs. In some tissues, addition of glycogen precipitates macromolecules like carbohydrates to remove these impurities; conventional methods can be combined with commercially available kits to obtain good-quality RNA.

6. RNA should be digested with DNase I to degrade any traces of DNA present in RNA and to ensure that no sequences originate from DNA during further processing and sequencing. Presence of DNA in RNA preparations contaminates sequencing data and may result in false predictions.

7. During loading of RNA onto a PAGE gel, thoroughly remove urea from well. Any traces of urea left in the well will lead to degradation of RNA.

8. Run times of PAGE gels should be standardized as the BPB dye moves differently in gels of different polyacrylamide percentages.

9. For using Novocraft, a reference genome is required to create an indexed reference genome. However, the genome sequence of any small bacterium or virus can be used here as it does not interfere with adapter removal. However, this will not work if you want to use other applications of the software.

10. In our case, SSAKE worked optimally with the parameters—m (minimum overlap) set at 16 and—t (end trimming) set to 2. Depending upon the data, these parameters may be varied.

11. Manuals of SSAKE, Velvet, and BLAST should be read carefully. One can try different options by adjusting parameters as suggested in the manuals and see the changes. Parameters with best results can be used for further analysis.

References

1. Jones R (2006) RNA silencing sheds light on the RNA world. PLoS Biol 4:e448
2. Vasudevan S, Tong Y, Steitz JA (2007) Switching from repression to activation: microRNAs can upregulate translation. Science 318:1931–1934
3. Wang XB, Jovel J, Udomporn P, Wang X, Wu Q, Li WX, Gasciolli V, Vaucheret H, Ding SW (2011) The 21-nucleotide, but not 22-nucleotide, viral secondary small interfering RNAs direct potent antiviral defense by two cooperative argonautes in *Arabidopsis thaliana*. Plant Cell 23:1625–1638
4. Almeida R, Allshire RC (2005) RNA silencing and genome regulation. Trends Cell Biol 15:251–258
5. Tomari Y, Zamore PD (2005) Perspective: machines for RNAi. Genes Dev 19:517–529
6. Chapman EJ, Carrington JC (2007) Specialization and evolution of endogenous small RNA pathways. Nat Rev Genet 8:884–896
7. Chen X (2010) Small RNAs—secrets and surprises of the genome. Plant J 61:941–958
8. Llave C, Kasschau KD, Rector MA, Carrington JC (2002) Endogenous and silencing associated small RNAs in plants. Plant Cell 14:1605–1619
9. Reinhart BJ, Weinstein EG, Rhoades MW, Bartel B, Bartel DP (2002) MicroRNAs in plants. Genes Dev 16:1616–1626
10. Sunkar R, Zhu JK (2004) Novel and stress-regulated microRNAs and other small RNAs from *Arabidopsis*. Plant Cell 16:2001–2019
11. Szittya G, Moxon S, Santos DM, Jing R, Fevereiro MP, Moulton V, Dalmay T (2008) High throughput sequencing of *Medicago truncatula* short RNAs identifies eight new miRNA families. BMC Genomics 9:593
12. Zhao CZ, Xia H, Frazier TP, Yao YY, Bi YP, Li AQ, Li MJ, Li CS, Zhang BH, Wang XJ (2010) Deep sequencing identifies novel and conserved microRNAs in peanuts (*Arachis hypogaea* L.). BMC Plant Biol 10:3
13. Ding SW, Voinnet O (2007) Antiviral immunity directed by small RNAs. Cell 130:413–426
14. Mlotshwa S, Pruss GJ, Vance V (2008) Small RNAs in viral infection and host defense. Trends Plant Sci 13:375–382
15. Donaire L, Wany Y, Gonzalez-Ibeas D, Mayer KF, Aranda MA, Llave C (2009) Deep sequencing of plant viral small RNAs reveals effective and widespread targeting of viral genomes. Virology 392:203–214
16. Kreuze JF, Perez A, Untiveros M, Quispe D, Fuentes S, Barker I, Simon R (2009) Complete viral genome sequence and discovery of novel viruses by deep sequencing of small RNAs: a generic method diagnosis, discovery and sequencing of viruses. Virology 388:1–7
17. Rwahnih MA, Daubert S, Golino D, Rowhani A (2009) Deep sequencing analysis of RNAs from a grapevine showing syrah decline symptoms reveals a multiple virus infection that includes a novel virus. Virology 387:395–401
18. Carra A, Mica E, Gambino G, Pindo M, Moser C, Pè ME, Schubert A (2009) Cloning and characterization of small non-coding RNAs from grape. Plant J 59:750–763
19. Zhang Y, Singh K, Kaur R, Qiu W (2011) Association of a novel DNA virus with the grapevine vein-clearing and vine decline syndrome. Phytopathology. doi:10.1094/PHYTO-02-11-0034
20. Sambrook J, Fritsch EF, Maniatis T (1989) Molecular cloning: a laboratory manual. Cold Spring Harbor Press, Cold Spring Harbor
21. Ghawana S, Paul A, Kumar H, Kumar A, Singh H, Bhardwaj PK, Rani A, Singh RS, Raizada J, Singh K, Kumar S (2011) An RNA isolation system for plant tissues rich in secondary metabolites. BMC Res Notes 4:85
22. Hafner M, Landgraf P, Ludwig J, Rice A, Ojo T, Lin C, Tuschl T (2008) Identification of microRNAs and other small regulatory RNAs using cDNA library sequencing. Methods 44:3–12
23. Altschul SF, Madden TL, Schäffer AA, Zhang J, Zhang Z, Miller W, Lipman DJ (1997) Gapped BLAST and PSI-BLAST: a new generation of protein database search programs. Nucleic Acids Res 25:3389–3402
24. Warren RL, Sutton GG, Jones SJM, Holt RA (2007) Assembling millions of short DNA sequences using SSAKE. Bioinformatics 23:500–501
25. Zerbino DR, Birney E (2008) Velvet: algorithms for de novo short read assembly using de Bruijn graphs. Genome Res 18:821–829

Chapter 15

Global Assembly of Expressed Sequence Tags

Foo Cheung

Abstract

The method for the construction of Expressed Sequence Tag (EST) assemblies described here uses reads generated from 454 pyrosequencing and Sanger and Illumina (Solexa) sequencing technologies as input. It is consistent with and parallels many established EST assembly protocols, for example the TIGR Gene Indices. Reads that are used as input to the EST assembly process usually come from both internal and external sources. Thus, in addition to internally generated EST reads, expressed transcripts are collected from dbEST and also the NCBI GenBank nucleotide database (full-length and partial cDNAs). "Virtual" transcript sequences derived from whole genome annotation projects can be excluded, depending on the needs of the project. Currently, in most cases, 454-derived sequences can be treated similar to Sanger-derived ESTs. In contrast, the shorter Solexa-derived sequences will have to undergo a round of either de novo assembly or an "align-then-assemble" approach against a reference genome, if available, before these transcripts can be used for the purpose of a global EST assembly that combines a mixture of Sanger and next-generation sequencing technologies.

Key words: Expressed Sequenced Tags, Sequencing, EST assembly, 454 pyrosequencing, Sanger, Illumina, Solexa

1. Introduction

Next-generation sequencing technologies (NGSTs) generate huge numbers of short read fragments in a fraction of the time and cost compared to Sanger sequencing. These technologies include Roche/454 pyrosequencing (http://www.454.com), Illumina SBS technology (http://www.illumina.com), and sequencing by ligation (http://www3.appliedbiosystems.com). We and others have found that deep sequencing of Expressed Sequence Tags (ESTs), made cost-effective by pyrosequencing technology, leads to significantly enhanced new gene discovery rates in a normalized cDNA library, and the short reads produced are of value in gene structure annotation (1, 2). ESTs are of great value because they provide evidence of expression of predicted genes, and by spliced alignment to genomic DNA they can provide support for gene structures. In instances where a genome sequence is not available, EST sequencing provides a first catalog of a species' gene inventory. Combinations of library normalization and deep sequencing are used to maximize gene discovery. ESTs are single-pass and partial sequences from cDNA clones that provide a rapid and cost-effective method to analyze transcribed portions of the genome while avoiding the noncoding and repetitive DNA that can make up much of the genome. EST sequencing has been shown to accelerate gene discovery, including gene family identification (3), large-scale expression analysis (4, 5), establishing phylogenetic relationships (6), developing PCR-based molecular markers (7), and identifying simple sequence repeats (8) and single-nucleotide polymorphisms (SNPs) (9). Previous studies have shown that ESTs derived from a single technology can be used to identify SNPs using Sanger (10) or 454 pyrosequencing technologies alone (11). We have extended this and have shown that a global assembly of ESTs approach can be used that combines Sanger ESTs and de novo-assembled Illumina sequences to correctly identify polymorphisms and transcript profiling in alfalfa, a non-model, allogamous, autotetraploid species (12). Similarly, we have also shown that ESTs derived from Sanger and 454 pyrosequencing technologies can be merged and then assembled to generate the *Pythium ultimum* transcriptome (13) using a global EST assembly approach.

2. Materials

The pipeline to assemble ESTs is normally carried out using high-performance computing, summarized in Fig. 1, and is consistent with other established EST assembly protocols, for example the TIGR Gene Indices (14) and the TIGR Plant Transcript Assemblies (15). Assemby of large EST datasets can be tested and carried out on virtualized, or "cloud," computing services, such as the Amazon Elastic Compute Cloud (EC2) (http://aws.amazon.com/ec2/), which provides an economical means to procure vast computing resources to facilitate parallel computation on an as-needed basis.

3. Methods

3.1. Collecting Public ESTs

The aim of this step is to increase the number of reads by combining both internally generated EST reads with any useful public EST datasets, which would normally include high-quality

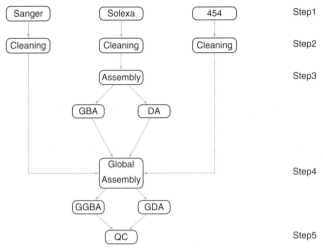

Fig. 1. Global EST assembly. Schematic representation of the steps and different sets of sequences involved in the global assembly of ESTs into contigs.

Sanger-derived EST data (16). Combining datasets increases contig depth, contig length, and the number of alternative splice variants. In many cases, computationally derived or predicted transcripts from genome annotation projects should be excluded so that any given sequence represents experimental proof that such sequence is actually transcribed. Using the following html link from the NCBI Entrez Taxonomy Homepage (http://www.ncbi.nlm.nih.gov/sites/entrez?db=taxonomy), EST and transcripts from a wide range of organisms can be queried and downloaded in fasta format. Additional NGS datasets can be downloaded from the Short Read Archive (http://www.ncbi.nlm.nih.gov/sra). The usefulness of adding additional datasets from this archive should be carefully examined on a case-by-case basis.

3.2. Cleaning ESTs

The aim of this step is to remove or trim a variety of reads that could interfere with assembly prior to further analyses. Reads will require trimming of low-complexity regions, poly-A and poly-T tracts, and sequence ends rich in undetermined bases, as well as low-quality sequences. A number of tools are available to help with this, including TIGR SeqClean tool (14), Vmatch (http://www.vmatch.de/), Crossmatch (http://www.phrap.org/phredphrap/phrap.html), SeqTrim (17), CLC-bio (http://www.clcbio.com/), emboss (18), and galaxy (19) (http://main.g2.bx.psu.edu/). Galaxy, SeqTrim, and SeqClean can identify sequence inserts, remove vector sequences from the NCBI UniVec database (http://www.ncbi.nlm.nh.gov/VecScreen/UniVec.html), low-quality sequences, adaptors, and low-complexity, poly-A/T, and contaminant sequences, and detect chimeric reads. SeqTrim and Galaxy can clean any type of fasta sequence reads and can be run via a Web interface (http://www.scbi.uma.es/cgi-bin/seqtrim/seqtrim_form.cgi) for several thousand reads or by command line for a larger number of reads. Another similar tool to SeqTrim is the TIGR SeqClean tool (http://www.tigr.org/tdb/tgi/software), which is a stand-alone program that takes in fasta sequence reads. After running blastn searches against a database of vectors, adaptors, and linkers, the outputs are the trimmed cleaned sequences by using the following command line:

$path/seqclean sanger.fasta -v $dbpath/UniVec, $dbpath/adaptors, $dbpath/linkers -o cleaned.transcripts.fasta

In addition, reads that have blastn matches to retrotransposons, mitochondria, and chloroplasts should also be filtered, as should sequences containing long simple sequence repeats and those sequences not passing stringent quality thresholds.

3.3. Assembly of Short Reads (<100 Nucleotides)

The aim of this step is to assemble the short reads generated from Illumina (Solexa) sequencing before proceeding to the global assembly. There are at least two strategies to reconstruct transcripts from short RNA-Seq reads. If a reference genome is available, then one can use an "align-then-assemble" approach. Again, there are numerous tools to tackle this, including running Seqman Ngen (DNAstar, Inc.) or the triad bowtie, tophat, and cufflinks (20–22) or CLC-Bio. Reads that did not join into contigs in the reference-based assembly should undergo de novo assembly. The resulting contigs and remaining singletons should then be combined into a single set. There are numerous tools for performing de novo assembly, including velvet (23), Trinity (24), Seqman Ngen (DNAstar, Inc.), CLC-bio, and Oases (http://www.ebi.ac.uk/~zerbino/oases/).

3.4. Assembly of ESTs

3.4.1. Global Genome Assembly of ESTs— "Align-Then-Assemble"

The aim of this step is to assemble ESTs based upon the genome backbone. However, if a reference genome is not available, skip this step and go to de novo assembly of ESTs. The Program to Assemble Spliced Alignments (PASA) can be used to create genome-based EST assemblies (25). The transcript alignments are clustered based on genome mapping location and assembled into gene structures that include the maximal number of compatible transcript alignments generated by PASA maximal alignment assemblies. Web-based tools are the method of choice to launch the pipeline, although command-line utilities are also available. Running PASA to generate alignment assemblies takes two inputs: a multi-fasta file for the genome and a multi-fasta file for the transcripts. Optionally, an additional file containing full-length cDNAs can be provided.

3.4.2. De Novo Global Assembly of ESTs

The aim of this step is to carry out de novo global assembly of ESTs when a reference genome is not available, or from reads that did not align to the genome sequence used in the previous step, presumably because the reference genome is incomplete. The resulting contigs and remaining singletons from both the genome and the de novo assembly are then combined into a single set. Numerous tools are available that can be used to assemble ESTs, for example Seqman Ngen (DNAstar, Inc.), paracel transcript assembler (http://www.paracel.com/), CAP3 (26), and CLC-Bio. Typically, we would use the TIGR Gene Index transcript clustering and assembly pipeline (14) when no reference genome is available and reads are above 100 bp. This pipeline is driven by the TGICL utility. Megablast (27) is first used to identify pairwise alignments between individual transcript sequences. Single-linkage clustering is used to generate clusters of matching transcripts. Each transcript cluster is fed to the CAP3 (64 bit) or preferably the Paracel Transcript Assembler. Outputs from this process include the assembled transcript sequences (tentative consensus sequences) and singletons (unassembled transcripts). The TGICL has been successfully used for creating EST assemblies. The TGICL tool, Megablast, and CAP3 assembler are used for the clustering and assembly steps using the following command line:

$path/tgicl $fasta -p 95 -150 -v 20 -s 10000 -O '-p 95 -y 20 -o 50'

TGICL is a wrapper script which first clusters the vector-trimmed sequences based on an all-versus-all pairwise sequence comparison using Megablast, and subsequently assembles each cluster using CAP3 or, in the case of large datasets, using the Paracel Transcript Assembler with the following highly stringent criteria: (1) the overlap between two sequences must be longer than 50 bp; (2) the sequence identity between the overlapping region of the two sequences must be >95%; (3) the number of overhanging, unaligned bases at the ends of the sequences must be <20 bases; (4) large clusters with >10,000 component sequences are avoided.

3.5. Quality Control

This step conducts some quality control analysis of the assembly process. Potential errors are indicated by the presence of alignment fragmentation to related genomes, cDNAs or ESTs, exon non-colinearity of gene models, or unexpectedly high read depths. Additional problems that can arise include Blastx hits to retroelements. These can be identified from protein matches that contain the terms "copia," "gag," "pol," "retroelement," "integrase," "reverse transcriptase," and "retrotransposon" in their annotation. In contrast, accurately assembled contigs are characterized by a high proportion of contigs containing matches to known proteins using BLAST searches, and by the experimental PCR amplification of SSR markers developed in contigs. Singletons having matches to proteins in BLAST searches indicate that they also provide an important source of information.

Acknowledgments

The author wishes to acknowledge funding and support from the NIH/NHLBI Center for Human Immunology.

References

1. Cheung F, Haas B, Goldberg S, May G, Xiao Y, Town C (2006) Sequencing *Medicago truncatula* expressed sequenced tags using 454 Life Sciences technology. BMC Genomics 7:272
2. Bourdon V, Naef F, Rao P, Reuter V, Mok S, Bosl G, Koul S, Murty V, Kucherlapati R, Chaganti R (2002) Genomic and expression analysis of the 12p11-p12 amplicon using EST arrays identifies two novel amplified and over-expressed genes. Cancer Res 62:6218–6223
3. Ewing R, Ben Kahla A, Poirot O, Lopez F, Audic S, Claverie J (1999) Large-scale statistical analyses of rice ESTs reveal correlated patterns of gene expression. Genome Res 9:950–959
4. Samuel Yang S, Cheung F, Lee J, Ha M, Wei N, Sze S, Stelly D, Thaxton P, Triplett B, Town C, Jeffrey Chen Z (2006) Accumulation of genome-specific transcripts, transcription factors and phytohormonal regulators during early stages of fiber cell development in allotetraploid cotton. Plant J 47:761–775
5. Nishiyama T, Fujita T, Shin-I T, Seki M, Nishide H, Uchiyama I, Kamiya A, Carninci P, Hayashizaki Y, Shinozaki K, Kohara Y, Hasebe M (2003) Comparative genomics of *Physcomitrella patens* gametophytic transcriptome and *Arabidopsis thaliana*: implication for land plant evolution. Proc Natl Acad Sci USA 100:8007–8012
6. Gupta P, Rustgi S (2004) Molecular markers from the transcribed/expressed region of the genome in higher plants. Funct Integr Genomics 4:139–162
7. Mian M, Saha M, Hopkins A, Wang Z (2005) Use of tall fescue EST-SSR markers in phylogenetic analysis of cool-season forage grasses. Genome 48:637–647
8. Rafalski A (2002) Applications of single nucleotide polymorphisms in crop genetics. Curr Opin Plant Biol 5:94–100
9. Varshney R, Thiel T, Stein N, Langridge P, Graner A (2002) In silico analysis on frequency and distribution of microsatellites in ESTs of some cereal species. Cell Mol Biol Lett 7:537–546
10. Kuhl JC, Cheung F, Yuan Q, Martin W, Zewdie Y, McCallum J, Catanach A, Rutherford P, Sink KC, Jenderek M, Prince JP, Town CD, Havey MJ (2004) A unique set of 11,008 onion expressed sequence tags reveals expressed sequence and genomic differences between the monocot orders Asparagales and Poales. Plant Cell 16:114–125
11. Han Y, Kang Y, Torres-Jerez I, Cheung F, Town CD, Zhao PX, Udvardi MK, Monteros MJ (2011) Genome-wide SNP discovery in tetraploid alfalfa using 454 sequencing and high resolution melting analysis. BMC Genomics 12:350
12. Yang S, Tu ZJ, Cheung F, Xu WW, Lamb JF, Jung HJ, Vance CP, Gronwald JW (2011) Using RNA-Seq for gene identification, polymorphism detection and transcript profiling in two alfalfa genotypes with divergent cell wall composition in stems. BMC Genomics 12:199
13. Cheung F, Win J, Lang J, Hamilton J, Vuong H, Leach J, Kamoun S, André Lévesque C, Tisserat N, Buell C (2008) Analysis of the *Pythium ultimum* transcriptome using Sanger and Pyrosequencing approaches. BMC Genomics 9:542
14. Pertea G, Huang X, Liang F, Antonescu V, Sultana R, Karamycheva S, Lee Y, White J, Cheung F, Parvizi B, Tsai J, Quackenbush J (2003) TIGR Gene Indices clustering tools (TGICL): a software system for fast clustering of large EST datasets. Bioinformatics 19:651–652
15. Childs K, Hamilton J, Zhu W, Ly E, Cheung F, Wu H, Rabinowicz P, Town C, Buell C, Chan A (2007) The TIGR Plant Transcript

Assemblies database. Nucleic Acids Res 35:D846–D851

16. Boguski M, Lowe T, Tolstoshev C (1993) dbEST-database for "expressed sequence tags". Nat Genet 4:332–333

17. Falgueras J, Lara A, Fernández-Pozo N, Cantón F, Pérez-Trabado G, Claros M (2010) SeqTrim: a high-throughput pipeline for preprocessing any type of sequence read. BMC Bioinformatics 11:38

18. Rice P, Longden I, Bleasby A (2000) EMBOSS: the European Molecular Biology Open Software Suite. Trends Genet 16:276–277

19. Goecks J, Nekrutenko A, Taylor J, Team G (2010) Galaxy: a comprehensive approach for supporting accessible, reproducible, and transparent computational research in the life sciences. Genome Biol 11:R86

20. Langmead B, Trapnell C, Pop M, Salzberg S (2009) Ultrafast and memory-efficient alignment of short DNA sequences to the human genome. Genome Biol 10:R25

21. Trapnell C, Pachter L, Salzberg S (2009) TopHat: discovering splice junctions with RNA-Seq. Bioinformatics 25:1105–1111

22. Trapnell C, Williams B, Pertea G, Mortazavi A, Kwan G, van Baren M, Salzberg S, Wold B, Pachter L (2010) Transcript assembly and quantification by RNA-Seq reveals unannotated transcripts and isoform switching during cell differentiation. Nat Biotechnol 28:511–515

23. Zerbino D, Birney E (2008) Velvet: algorithms for de novo short read assembly using de Bruijn graphs. Genome Res 18:821–829

24. Grabherr MG, Haas BJ, Yassour M, Levin JZ, Thompson DA, Amit I, Adiconis X, Fan L, Raychowdhury R, Zeng Q, Chen Z, Mauceli E, Hacohen N, Gnirke A, Rhind N, di Palma F, Birren BW, Nusbaum C, Lindblad-Toh K, Friedman N, Regev A (2011) Full-length transcriptome assembly from RNA-Seq data without a reference genome. Nat Biotechnol 29:644–652

25. Haas B, Delcher A, Mount S, Wortman J, Smith RJ, Hannick L, Maiti R, Ronning C, Rusch D, Town C, Salzberg S, White O (2003) Improving the Arabidopsis genome annotation using maximal transcript alignment assemblies. Nucleic Acids Res 31:5654–5666

26. Huang X, Madan A (1999) CAP3: a DNA sequence assembly program. Genome Res 9:868–877

27. Zhang Z, Schwartz S, Wagner L, Miller W (2000) A greedy algorithm for aligning DNA sequences. J Comput Biol 7:203–214

Chapter 16

Computational Analysis of RNA-seq

Scott A. Givan, Christopher A. Bottoms, and William G. Spollen

Abstract

Using High-Throughput DNA Sequencing (HTS) to examine gene expression is rapidly becoming a viable choice and is typically referred to as RNA-seq. Often the depth and breadth of coverage of RNA-seq data can exceed what is achievable using microarrays. However, the strengths of RNA-seq are often its greatest weaknesses. Accurately and comprehensively mapping millions of relatively short reads to a reference genome sequence can require not only specialized software, but also more structured and automated procedures to manage, analyze, and visualize the data. Additionally, the computational hardware required to efficiently process and store the data can be a necessary and often-overlooked component of a research plan. We discuss several aspects of the computational analysis of RNA-seq, including file management and data quality control, analysis, and visualization. We provide a framework for a standard nomenclature system that can facilitate automation and the ability to track data provenance. Finally, we provide a general workflow of the computational analysis of RNA-seq and a downloadable package of scripts to automate the processing.

Key words: High-throughput DNA sequencing, RNA-seq, Gene expression, Data processing

1. Introduction

High-Throughput DNA Sequencing (HTS) is rapidly becoming the preferred choice to determine the qualitative and quantitative characteristics of a sequence library. When applied to a library derived from expressed sequences, it is typically referred to as "RNA-seq." As HTS technologies become more readily available to researchers, RNA-seq is poised to overtake microarrays, the predominant technology of the last decade for whole genome expression analysis. Just as the early days of microarray analysis were marked by unprecedented data breadth, nonstandardized algorithms, and file formats, there are a variety of unfamiliar decisions and alternatives facing a researcher embarking on an RNA-seq analysis. It is beyond the scope of this chapter to delve into an in-depth comparison of the various software packages for RNAseq. And, considering the rapid development cycle, any software evaluation provided now would quickly be out of date. Rather, we present a set of best practices for working with HTS data with particular attention given to RNA-seq. Additionally, we describe the RNA-seq Toolkit (RST), a collection of scripts that automate the use of contemporaneous software the authors use for RNA-seq analysis: the FASTX-Toolkit (1), SAMtools (2) and the "Tuxedo" software suite [Bowtie (3), TopHat (4) and Cufflinks (5)]. Although alternatives exist for several programs in the pipeline, we hope that discussing the specifics of our procedures will provide guidance within the broader scope of RNA-seq software packages and facilitate the evaluation of efficacy of other software packages. Finally, all tools discussed in this article apply to RNA-seq analyses for which a reference genome exists. For RNA-seq analysis using data for which no reference genome exists, please refer to the chapter by Bryant and Mockler.

2. Materials

We recommend that anyone undertaking an analysis of HTS data have some experience working on the command line within the Linux operating system. If not, we suggest reading through one of a plethora of online tutorials; for example, http://tldp.org/LDP/intro-linux/html/index.html. Becoming familiar with several basic Linux commands will be necessary to mimic the steps described within this manuscript; for example, commands like mkdir, cd, mv, rm, cat, ln, ls, and less. Documentation for these commands can be found in the "man" pages on any Linux machine. For example, to learn about the command "mkdir" after logging into a Linux machine, invoke the command "man mkdir." Alternatively, Web pages derived from the "man" pages can be identified using Internet search engines.

All of the scripts and programs discussed below are designed to work on a computer using the Linux operating system (see Note 1). This includes operating systems similar to Linux, including Unix and Macintosh OS X. We verified that RST works properly on several specific Linux distributions; including Fedora release 13, Cent OS 5.5 (Final), Red Hat Enterprise Server (RHEL) release 5.5 (Tikanga), and Ubuntu release 10.10. We verified that RST functions properly on Macintosh OS 10.6.5. Finally, we verified that RST works properly on Bio-Linux release 6. However, several additional software packages or libraries are required, which may necessitate separate installation procedures depending on the specific platform and operating system. These include: SAMtools (2), version 0.1.8; FASTX-Toolkit (1), version 0.0.13; Bowtie (3), version 0.12.7; TopHat (4), version 1.1.4; Cufflinks (5), version 0.9.3; and Perl, version 5.8.8. We have summarized all dependencies, with links to their download locations, here: http://ircf.rnet.missouri.edu:8000/tech/illumina/dna_sequencing/rnaseq/tophat_pipeline/dependencies or http://ow.ly/3wm4B.

We have developed a set of scripts to automate running a RNA-seq pipeline, which we refer to as the RNA-seq Toolkit or RST. A compressed archive containing all scripts associated with RST and sample data can be downloaded from this URL: http://ircf.rnet.missouri.edu:8000/tech/illumina/dna_sequencing/rnaseq/tophat_pipeline/toolkit/ or http://ow.ly/45WZf.

3. Methods

3.1. Data Management

It is important to establish a management plan for data associated with an RNA-seq project. A management plan should contain procedures related to at least these four aspects: data retrieval, short-term storage, data backup and long-term or archival storage. There will likely be project-specific characteristics that bear on each of these aspects, but we will provide some general ideas and recommendations. See Note 2 for a discussion of disk space recommendations.

Data retrieval is often dictated by the laboratory or facility that generates the HTS data. Researchers should request specific guidelines regarding delivery methods for the data and approximate disk space requirements. Unless specialized hardware is in place, it is often infeasible to transfer entire datasets across computer networks that span a large physical distance; for example, more than several hundred miles. Therefore, it is relatively common to use courier-based methods to transfer computer hard drives between data generation and data analysis locations. Other possible data transfer solutions include the FASP file transfer protocol (http://www.ncbi.nlm.nih.gov/books/NBK47532/#SRA_Submission_Guid.I5331_The_fasp_Protoorhttp://ow.ly/45X4a) or BioTorrents (6).

Short-term storage is used during the data analysis stages of a project and is often of very high-performance with high data transfer capabilities. The details of the short-term storage hardware usually depend on the existing institutional computer infrastructure. University researchers in the United States frequently have access to a computational infrastructure dedicated to research purposes. Often these resources will include large storage networks that can be used during short-term analysis timeframes (less than 1 year). These typically include redundant disk systems (for example,

Redundant Array of Independent Disks or RAID) that contain dozens of disks and serve as an excellent foundation to facilitate data analysis. If these resources are unavailable, it is possible to build smaller redundant disk systems using relatively inexpensive commodity hardware available from several retailers. Look for a 4 or 5 bay RAID enclosure with a modern I/O interface such as USB 2.0, Firewire 800 (1394b), or eSATA.

An often-overlooked component of a robust data management plan is a backup strategy. Most well-managed computational infrastructures will offer some type of data backup service. Ideally, data backup will be automated and potentially involve both on-site and off-site resources. On-site backups could be simply an additional copy of the data on the same or different file system or an archive copied to digital tape, if available. On a Linux computer, rsync is a popular and effective program that can maintain copies of a directory structure on a local or remote disk system. rsync can be invoked to run in "archive" mode, which, in general, only copies changed or new files to a backup location. For more information about rsync, users should consult the rsync man page ("man rsync"), or http://rsync.samba.org/. An interesting alternative is to use one of the several "cloud"-based backup systems available from online commercial vendors; for example, Mozy, http://mozy.com/, ADrive, http://www.adrive.com, and Memopal, http://www.memopal.com. Although cost and the limited speed of network transfers may be a consideration, these solutions theoretically provide extremely robust long-term data backup. A particular limitation for many of these vendors is that their software is restricted for use on Microsoft Windows-based computers. There may be a Macintosh client, but rarely one for Linux. One way to solve this is to choose a vendor who offers a WebDAV interface, which can be mapped as a network-attached file system, potentially to any type of computer. Of the companies listed, above, ADrive offers WebDAV and Memopal offers a (beta) Linux client.

In the context of HTS data, archival storage refers to storing data that is unlikely to be altered, is typically "read-only" and meant to be available via a publicly accessible interface. Although the US National Center for Biotechnology Information (NCBI) recently announced the phase out of specific parts of the Sequence Read Archive (SRA), researchers can continue to submit short read sequences to the European Nucleotide Archive (ENA) (http://www.ebi.ac.uk/ena/) or DNA Databank of Japan (DDBJ) (http://www.ddbj.nig.ac.jp/), and processed RNA-seq data to the NCBI Gene Expression Omnibus (GEO) (7) or EMBL-EBI ArrayExpress (8).

To facilitate automated analyses, it is highly advantageous to develop standard naming conventions for the directories and files containing the data and results. The scripts included in RST use a convention developed in our group that enforces three general principles; for details, see Standardized Nomenclatures, below. First, use file and directory names that are meaningful and, when possible and advantageous, names should reflect the data source. Second, names should reflect actions taken on the data. Third, use nondescriptive or generic symbolic link names to point to data files conforming to the first two naming conventions. Note that developing and using a script to automate any analysis tends to enforce naming conventions, no matter how ad hoc they may be. Additionally, developing a convention for organizing and naming files and directories will greatly facilitate any future automation.

3.2. Quality Control of Sequence Reads

In general, we trim or remove sequences based on two characteristics: low base quality confidence scores and similarity to Sequences that Originate from Dubious sources (SOD). As discussed below, low-quality bases can be generated for a number of reasons and can be removed using a number of methods. Likewise, SODs may arise for a number of reasons and can be removed based on their similarity to a set of sequences collected to represent likely SODs for a particular dataset, which is often dictated by the organism from which the bulk of the sequence reads should originate, as discussed below.

Base calling on any of the commonly used platforms can be compromised by low-quality starting material, errors in chemistry or faults in the instrument. Prior to any alignment or assembly of the reads, one should remove low-quality bases. Given the large number of reads generated by high-throughput sequencing and the improving quality of the base calls with refinements in chemistry, hardware, and software, this usually means that many sequences remain even after rather aggressive Quality Control (QC) measures. But even if this is not the case, it is advantageous to work with a clean dataset since incorrect reads slow down subsequent processing, especially during de novo assembly where potential sequencing errors can greatly increase the run time and memory needed.

Errors in the identification of a base may arise from platform-specific steps, e.g., when using the Illumina technology several clusters substantially overlap or bubbles occur in the lane of the flow cell, or, in the SOLiD system, when there is more than one clone on a bead. Developments in hardware, chemistry, and software have lessened some of these problems, but it is still advantageous to identify and remove low-quality sequence prior to analysis.

The quality of an Illumina or SOLiD base call tends to be inversely correlated to its position in the extending DNA strand; for example, base quality confidence progressively decreases 5′ to 3′. Therefore, bases that fall below some quality confidence threshold are usually removed from the 3′ end of a read. The read is retained if the remaining sequence is long enough and contains a sufficient number of high-quality bases. The degree of confidence for a base is usually quantified as a phred score (9). RST uses two programs included in the FASTX-Toolkit (1), fastq_quality_trimmer and fastq_quality_filter, for trimming from the 3′ end and filtering sequences based on windows of average base quality, respectively, as determined by a threshold phred score (see steps 6a and 6b in the Subheading 3.4.2, below). Typically that threshold is less stringent for sequence alignment than for sequence assembly.

Once low-quality bases or reads are culled, SODs can be removed. Typical SODs in RNA preps are highly similar to rRNA or DNA from plastids or mitochondria. Although likely less of a concern for RNA-seq experiments than for genome assemblies, when working with nonhuman samples it may be valuable to identify and remove SODs highly similar to the human AluY repeat element, ostensibly caused by trace contamination events (10). Also, anecdotal reports suggest that a common class of SOD seemingly originates from the bacteriophage PhiX genome. PhiX DNA is used as a control sample on the Illumina platform. It is unclear whether these reports of contamination originate from the lab or are due to actual sequence similarity between PhiX and the samples being sequenced. In a review of experiments using the Illumina platform at the DNA Core at the University of Missouri, we found no evidence of leakage of the PhiX sample to adjacent lanes, nor was there any reason to suspect lab contamination (Nathan Bivens, University of Missouri DNA Core, personal communication). Thus, it is still unclear whether PhiX DNA truly is a common contaminant.

Although several comparably effective options exist, we usually use the program bowtie (3) to filter HTS reads against a file of representative SODs. To facilitate filtering, we exploit the --un command line option, which causes bowtie to output all the sequences that fail to align to the SODs (see step 6c in the Subheading 3.4.2, below). Effectively, this generates a FASTQ file to be used for further processing. Readers are encouraged to consult the bowtie manual (http://bowtie-bio.sourceforge.net/manual.shtml) for further discussion of its use (see Note 3).

Finally, keeping record of the numbers of sequence reads passing each step of trimming and filtering helps to identify inefficiencies or abnormalities in the processing or generation of the datasets.

3.3. Genome Masking

Using RepeatMasker (11), it is possible to mask tracts of genome sequence based on their similarity to known repetitive elements. In some genomes, for example, *Zea mays*, these masked regions can comprise over 85% of the genome (12). Thus, it can be useful to mask genome sequences to focus the alignment of RNA-seq reads towards the regions of the genome that likely contain the majority of expressed elements. However, considering that repetitive elements can lie within gene-rich regions (13) and that the expression of repetitive elements can correlate with certain biological functions or developmental phases (14), we typically do not mask a genome sequence prior to its use by an aligner like bowtie. Careful consideration of the implications of masking a genome sequence is recommended.

3.4. Data Analysis

3.4.1. Standardized Nomenclatures

Before proceeding to the general workflow, below, it is advantageous to discuss specific aspects of directories and files that can aid the organization and management of the data. As discussed, above, often the quantity of data generated during an RNA-seq project dictates the use of multiple file systems. Thus, it can be useful to use file system symbolic links to group the datasets logically into a directory structure that reflects the biological aspects of the experiment. For example, as displayed in Fig. 1a, six files may contain the paired-end sequence data from three samples and reside within three different directories on three different file systems (many more complicated scenarios are possible). The nomenclature presented is typical of Illumina output, where the file name s_3_1_sequence.txt and s_3_2_sequence.txt represent the paired reads from lane 3 of the flow cell (in this case, flow cell 218). As illustrated, these data correspond to sample "s2" of the RNA-seq experiment, which includes two other paired-end samples, s1 and s3. As illustrated in Fig. 1b, it is convenient to create a single data analysis directory and group the different samples in their own directories within. Since the actual data files reside on different file systems, we generally create symbolic links within the sample directories that point to the files. Thus, from within the directory "all_data," it is trivial to run commands on all files included in the experiment. For example, the last two lines of Fig. 1c are commands to validate the integrity of some types of fastq files. To run the pipeline, described below, a similar file and directory architecture are required.

Another practice is to create a formal system of file nomenclature. There are several pipeline steps that create output files that are used directly by subsequent steps and may be of use for other purposes. We recommend a nomenclature system that includes information about all the actions taken on a file, which enables a history of the actions to be derived solely from the file name. For example, as illustrated in Fig. 2, an original input file may be named set1.fq. The output file from the program fastq_quality_trimmer is named set1_qt.fq; the "qt" in the file name represents that the sequences have been *q*uality-*t*rimmed. set1_qt.fq serves as the input for the next program, fastq_quality_filter, which generates an output file named set1_qt_qf.fq, where "qf" represents that the sequences have been *q*uality-*f*iltered. Finally, bowtie is used to filter the sequences based on their similarity to SODs and the output file prefix is

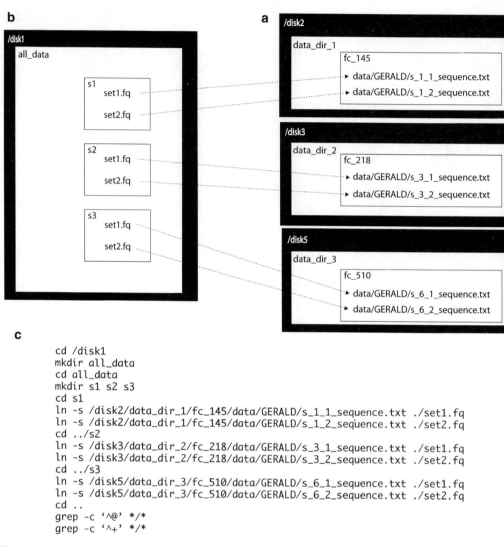

Fig. 1. Example of typical RNA-seq directories, data files, and symbolic links. (a) Six data files (s_1_1_sequence.txt, s_1_2_sequence.txt, s_3_1_sequence.txt, s_3_2_sequence.txt, s_6_1_sequence.txt and s_6_2_sequence.txt) reside on three different file systems. /disk1, /disk2, /disk3, /disk5 represent distinct file system volumes, which typically, but not necessarily, mean separate physical volumes. Note that, in this illustration, the full path of s_1_1_sequence.txt is /disk2/fc_145/data/GERALD/s_1_1_sequence.txt. /disk2 is the volume and fc_145, data, and GERALD are directories. This pattern applies to all the files. (b) /disk1 is a distinct file system volume that contains a typical analysis directory and contains symbolic links, represented by *dash lines*, to the data files on other distinct file systems. (c) The actual commands that could create the directory structures and symbolic links in (a) and (b).

appended with "sf" to represent that they have been subjected to *similarity-filtering*. Thus, each step of the process can be inferred from the final file name "set1_qt_qf_sf.fq." Finally, the "fq" suffix on all the file names, above, implies that they are in FASTQ format. Strict adherence to this, or a similar, system greatly increases the usability of output files generated by any step of the pipeline.

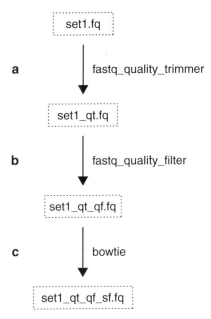

Fig. 2. Example of descriptive file nomenclature. File names are contained within *dotted boxes* and the programs acting on the data are represented by names next to *arrows*. The data in an input file, set1.fq, is progressively subjected to different steps in a typical RNA-seq QC pipeline (see text). Subsequent to the first step, (**a**), a new file contains sequences trimmed of low-confidence base calls. This file contains the "prefix" of the first file with "qt" appended to represent that these sequences have been quality trimmed. Likewise, in (**b**) and (**c**), output files are named to include actions taken on the data. For example, subsequent to (**b**), the file contains "qf" to represent *quality-filtering* and, subsequent to (**c**), the file name contains "sf" for *similarity-filtering*. The name of the final file reflects the original source of the data, set1.fq, and each action taken on the file.

3.4.2. General RNA-seq Workflow

Below, we present a general workflow of the computational steps of an RNA-seq analysis starting from FASTQ format sequence input files. The workflow is presented in outline form and includes example Linux commands to accomplish the steps. Typically, researchers first encounter the data at Step 3. As discussed in the following section, most of these steps can be automated using a scripted pipeline, such as RST. The names of the specific scripts within RST are included here within square brackets.

1. Generate Sequence Data from RNA-seq Library.
2. Transfer Data to Client.
3. Create Archive Copy of Data.
   (a) This copy will be used only to recreate original data in case of loss or hardware failure.
      i. cp -pr data_directory archive_directory.
   (b) Make all data files read-only.
      i. cd /path/to/data/directory.
      ii. chmod 0444 sequencefile*.txt.

4. Prepare Data Files.
   (a) Copy FASTQ files into place within defined directory scheme, or
      i. cp sequencefile1.txt /path/to/analysis/directory/set1.fq.
   (b) Create symbolic links to data files.
      i. ln -s /path/to/sequencefile1.txt set1.fq.
5. Validate Data Integrity of FASTQ files.
   (a) Count number of sequences in fastq file.
      i. Should know what to expect from sequencing run summary.
      ii. grep -c -P "^\@[\w:-]+?\:\w+?\s[12]\:.+" set1.fq.
   (b) Reconcile number of sequence lines and quality lines.
      i. number of sequences should equal number of quality strings.
      ii. grep -c -P "^\+([\w:-]+?\:\w+?\s[12]\:.+|\n)" set1.fq
   (c) If working with paired-end data contained within separate files, both files for a lane should have equal numbers of sequences and qualities.
6. Begin analysis: Quality Control [RNAseq_process_data.sh/ preprocess_fq.sh].
   (a) Trim low-quality bases from 3′ end of reads.
      i. fastq_quality_trimmer -t 13 -l 32 -i set1.fq -o set1_qt.fq.
   (b) Filter low-quality reads.
      i. fastq_quality_filter -p 90 -q 13 -o set1_qt_qf.fq.
   (c) Filter by sequence similarity.
      i. bowtie -q --solexa1.3-quals --un set1_qt_qf_sf.fq set1_qt_qf.fq.
7. Map RNA-seq reads to reference genome [RNAseq_process_data.sh / RNAseq.sh].
   (a) tophat (see below).
8. Assemble empirical transcripts [RNAseq_process_data.sh/ RNAseq.sh].
   (a) cufflinks (see below).
9. Identify high-value transcripts.
   (a) Filter by length [transcript_filter.pl].
   (b) Filter by coverage [cc_gtf.pl].
   (c) Occur in one or more samples with at least a minimal expression level (see --agg_transcripts option to RNAseq_process_data.sh, below).

　　　　i. Additional filtering can be facilitated by cc_gtf.pl, which filters by transcript coverage (see cufflinks documentation).

　　　　ii. cc_gtf.pl --cumulative --mincov 2.

10. Optional: re-run tophat using high-value transcripts, identified in step 9, as reference transcripts.

11. Estimate and identify potentially significant inter-sample transcript expression differences.

　　(a) cuffdiff (see below).

3.4.3. The RNA-seq Toolkit

As mentioned above, it is possible and often advantageous to automate an RNA-seq data analysis pipeline. The advantages are often realized through standardizing and codifying the procedure, which increases the consistency and robustness of the analysis across multiple samples or experiments. A scripted pipeline also facilitates the evaluation of changes and greatly simplifies debugging analysis problems, anomalies, and artifacts. We have developed an automated RNA-seq data analysis pipeline, which we refer to as the RNA-seq Toolkit (RST). We use RST regularly and it will be maintained into the foreseeable future. A bundled package of the scripts in RST is available for download, see above. RST is composed of Bash and Perl scripts and exploits the functionalities of the SAMtools (2) and FASTX-Toolkit (1) suites, as well as the "Tuxedo" suite [Bowtie (3), TopHat (4), Cufflinks (5)]. To work properly, RST requires these external software packages to be installed on the computer supporting the RNA-seq analysis.

The main wrapper script, RNAseq_process_data.sh, is the primary interface to RST. RNAseq_process_data.sh processes input data files and potentially analyzes an entire RNA-seq project from FASTQ data files through determination of transcript expression differences between samples. Command line options enable the actions of the script to be tailored for a particular data set, see Table 1. Default values are provided for most of these options, which allows RST to generate reasonable output for most datasets. To view all the available options and their default values, invoke the script as "RNAseq_process_data.sh -h."

RST is modular and specific functionalities can be included or excluded, as desired. Much of the modularity is embodied as distinct scripts that are invoked by RNAseq_process_data.sh. For example, as discussed above, passing the raw sequence reads through a series of QC steps can improve the robustness of an RNA-seq analysis. The QC functionalities are coded in the preprocess_fq.sh script. For paired-end data, when run in "full" mode using the --full flag, RST QC functionality invokes a four-step process. The first two steps use a pair of programs from the FASTX-Toolkit, fastq_quality_trimmer, and fastq_quality_filter to trim low-quality bases and filter entire reads of low quality, respectively.

Table 1
Command line options available for RNAseq_process_data.sh

Option	Alt. option	Default value	Explanation
--full	-f		Run QC, read mapping, data merging, transcript assembly
--partial	-p		Run mapping, data merging, transcript assembly
--transcripts	-a		Map reads according to transcript models in 'transcripts.gtf' input file
--mate_inner_distance	-r	165	Expected distance between mate pairs
--min_intron_length	-i	50	Minimum intron length
--max_intron_length	-I	10,000	Maximum intron length
--agg_transcripts	-t		Generate aggregate transcripts file
--refseq	-s	refseq	Name of reference sequence file
--threads	-H	8	Number of processors to use
--library_type	-l	fr-unstranded	Type of sequence library (see Tophat Web page)
--seonly	-e		Invoke if working with non-paired-end sequence data
--adapter	-A		Sequence to trim from all reads, if present
--indexpath	-P	index	Path to refseq and indicies
--help	-h		Display help menu

Many of these options and values are passed to other programs in or associated with RST. There are two ways to pass options to the script, indicated by the "Option" and "Alt. Option" columns. The single dash syntax is supported by most operating systems. If an option accepts a value, the default value is indicated in the "Default" column

The third step uses bowtie to filter out SODs, as described above. The fourth step uses a custom Perl script, fastq_pe_matchup.pl, described below. Although each of these steps is important for subsequent analysis, once run they do not need to be repeated, unless different parameters are desired. Thus, in subsequent runs RST can be invoked in "partial" mode using the --partial flag to not run the QC module, which can save a significant amount of time. Invoking RNAseq_process_data.sh in full and partial modes is illustrated in Fig. 3b and c, respectively.

As indicated in Fig. 3a, an additional step can be added to the QC process using the --adapter_seq flag. This flag should only be used when there are nucleotides common to all reads that should be removed before aligning to a reference genome. --adapter_seq requires as an argument the actual DNA sequence to be removed. Using this flag causes RNAseq_process_data.sh to run the adapter_trim.pl Perl script, which handles the sequence trimming manipulations. As an added QC functionality, enabled by default, any read

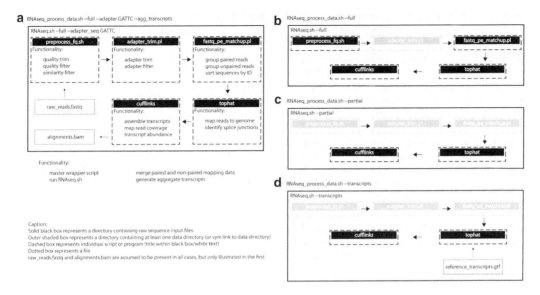

Fig. 3. Graphical representation of typical invocations of the script RNAseq_process_data.sh and its modular architecture. RNAseq_process_data.sh is a "wrapper" script that encompasses all functionalities of RST. Much of the functionalities are contained within modules, which are either scripts or binary executables. (**a**) As illustrated, the command invoked is "RNAseq_process_data.sh --full --adapter GATTC -agg_transcripts." This invokes another script as "RNAseq.sh --full --adapter GATTC." RNAseq.sh progressively calls five other programs: preprocess_fq.sh, a shell script that invokes programs associated with the QC pipeline, adapter_trim.pl, a Perl script included with RST that trims barcodes from reads, fastq_pe_matchup.pl, a Perl script included with RST that mates paired reads and segregates unmated reads to separate files, tophat, the main program associated with the TopHat suite, and cufflinks, a binary executable that assembles transcripts based on read alignments to a reference genome. The input file for the pipeline is represented by the *dotted box* named "raw_reads.fastq," whereas the output file is represented by the *dotted box* named "alignments.bam." Additionally, since --agg_transcripts was included in the invocation, an additional step will be run to assemble transcripts based on the alignments in alignments.bam. These data will be contained within a separate directory called "cufflinks" for each sample. The *solid black box* represents a directory containing raw_reads.fastq. The *outer shaded box* represents a directory containing at least one data directory or a symbolic link to a data directory. *Dashed boxes* represent individual scripts or programs, with their titles in white text within *black boxes* and their general functionalities described within. raw_reads.fastq and alignments.bam are assumed to be present in all cases, but only explicitly illustrated in the first. (**b**) When RNAseq_process_data.sh is invoked without --agg_transcripts and --adapter, adapter_trim.pl is not run and no transcripts are assembled. (**c**) When RNAseq_process_data.sh is invoked as --partial, preprocess_fq.sh is not run and the QC steps are skipped. This mode is intended to use sequences previously subjected to QC. (**d**) When RNAseq_process_data.sh is invoked with --transcripts, tophat uses a reference transcripts GTF file to only identify reads that align to those transcript models. This can be useful if a reliable source of annotation containing gene models is available for the reference genome.

that contains more than one copy of the "adapter" sequence is discarded. Our testing has demonstrated that the vast majority of reads containing more than one "adapter" sequence are composed of concatenated adapters. Therefore, fewer reads may be contained in the output file after processing with adapter_trim.pl.

When using paired-end data with the Tuxedo suite, the sequences within the two input FASTQ files must contain each pair member and the reads of each pair must be in the same order in both files. These requirements can be disrupted during QC steps and render the input files useless for further analysis using tophat. RST uses a custom Perl script, fastq_pe_matchup.pl, to group paired reads, group unpaired reads, sort sequences by their ID, and prepare the input files for tophat so that these requirements are fulfilled. fastq_pe_matchup.pl generates four files for a given pair of input FASTQ files. For a pair of input files named set1.fq and set2.fq, the output files will be set1.fq.matched.fq, set2.fq.matched.fq, set1.fq.nomate.fq, and set2.fq.nomate.fq. The two matched.fq files contain the properly paired and sorted sequences from set1.fq and set2.fq, while the two nomate.fq files contain the sequences from set1.fq and set2.fq that lack one member of the pair ("singles"), respectively. These four files will serve as input to the rest of the pipeline. When analyzing non-paired-end data, a single input file named set1.fq is required by RST.

Subsequent to the preprocessing steps, the four FASTQ files, described above, are used as two separate pairs of input files for tophat, which uses bowtie to align the reads to a reference genome and empirically maps potential splice junctions. The pair of matched.fq files is input for tophat running in paired-end mode. The pair of nomate.fq files is input for tophat running in non-paired-end mode. Thus, even though paired reads may be disrupted by a QC step, every read that passes the QC filters will be subjected to alignment to the reference genome by tophat.

RST generates a merged data file in BAM format containing all the paired-end and non-paired-end read alignments. RST uses samtools to generate the file, which is contained in the "merged" directory for each dataset. merged.bam can subsequently serve as input for cufflinks.

RST uses cufflinks to assemble transcripts from the alignments generated by tophat. Both the paired-end and non-paired-end reads are used in the assemblies to maximize the utility of the data. The output of cufflinks is contained in the "cufflinks" directory for each dataset.

To restrict an RNAseq analysis to only those transcripts that have been previously defined, use the --transcripts flag and an input file in GTF format with RNAseq_process_data.sh. This flag is conveyed to tophat and users should consult the tophat documentation (4) for details of what the GTF file should contain. The GTF file should be named or renamed transcripts.gtf and placed in the directory from which RNAseq_process_data.sh is invoked. Alternatively, create a symbolic link named transcripts.gtf pointing to the actual GTF file.

To build a GTF file of empirically derived transcript models, use the --agg_transcripts flag with RNAseq_process_data.sh, as illustrated in Fig. 3a. This flag causes the script to use cuffcompare to collect all the transcript models that occur at least once in any given dataset to generate a GTF file called transcripts.gtf in the directory from which RNAseq_process_data.sh was invoked. This file can be used with the --transcripts flag in subsequent analyses to direct alignments exclusively towards these gene models. Note that the transcripts GTF file can reflect alignment data that potentially span several datasets and could represent alignment data of an entire project. Organizing datasets by sample type or time point could limit the scope of --agg_transcripts to be more biologically meaningful.

3.4.4. Quantifying Transcript Levels

To facilitate the estimation of transcript levels in RNA-seq datasets, we frequently use the Cufflinks suite (cufflinks, cuffdiff, and cuffcompare). Although other options exist, and may be better suited for particular analyses, cufflinks is well integrated with the other components of the Tophat suite, which makes running the programs more straightforward. A typical invocation of cufflinks may look like this:

```
cufflinks\
--max-intron-length    10000    --min-intron-length    50
--reference-seq /path/to/refseq.fa \
--library-type fr-unstranded --label sample1 --GTF transcripts.gtf\
--num-threads 4 --output-dir cufflinks merged/merged.bam
```

Several of these options are tailored to the particular RNA-seq data set; for example: --max-intron-length, --min-intron-length, --library-type, --reference-seq, --label; whereas others are machine specific; for example: --num-threads. Other options, like --output-dir and --GTF, may be pertinent to all the datasets. Finally, the last argument (merged/merged.bam) is the path to a BAM or SAM file of RNA-seq alignment data, such as is generated by tophat (see Note 4). Cufflinks generates several output files, including a GTF file specifying the coordinates of the empirical transcript models.

When working with several samples, it is often useful to generate a global reference transcripts GTF file containing all the empirical transcripts represented across the experiment. Cuffcompare can be used to generate this reference file. As mentioned above, RST runs cuffcompare when the --agg_transcripts command line option is used.

3.4.5. Sample Comparisons

Subsequent to estimating transcript abundances within a single data set, generally the next step is to compare those estimates between different data sets. The Cufflinks suite includes the cuffdiff program to do these comparisons. As of Cufflinks release 0.9.0, cuffdiff facilitates the analysis of multiple data sets as either a time course or as technical or biological replicates, which significantly increased its utility. Invoking cuffdiff with an input of multiple data sets, which is most common for our group, generates output containing the evaluation of transcript differences between pairs of samples. Thus, some downstream processing of the output files may be necessary, but is usually rather trivial. A typical invocation of cuffdiff may be similar to:

```
cuffdiff \
--num-threads   4   --quartile-normalization   --output-dir
   cuffdiff_output
--reference-seq ../index/refseq.fa --library-type fr-unstranded \
--labels sample1,sample2,sample3 \
/path/to/reference/transcripts.gtf \
/path/to/sample1_0.bam,/path/to/sample1_1.bam,
/path/to/sample1_2.bam \
/path/to/sample2_0.bam,/path/to/sample2_1.bam \
/path/to/sample3_0.bam,/path/to/sample3_1.bam,
/path/to/sample3_2.bam \
```

Similar to the invocation of cufflinks, above, several of these options are tailored to the particular RNA-seq data set; for example: --reference-seq, --labels and --library-type; whereas others are machine specific; for example: --num-threads. Other options, like --output-dir and --quartile-normalization apply to all samples. After the dashed arguments, discussed above, the next argument specifies the location of a GTF file containing the coordinates of all the transcripts to analyze. Typically, this GTF file is generated by cufflinks, cuffcompare, or some other source. Finally, all subsequent arguments specify the locations of input files in BAM or SAM format, which contain the RNA-seq alignments for each sample to the reference genome. These arguments are space-delimited comma-separated lists: technical or biological replicates are grouped and separated by commas, while different sample types are delimited by spaces. Thus, in the example invocation, above, there are three groups of sample input files: samples one and three have three replicates, sample1_0.bam, sample1_1.bam, sample1_2.bam and sample3_0.bam, sample3_1.bam, sample3_2.bam, respectively, whereas sample two has two replicates, sample2_0.bam and sample2_1.bam.

3.5. Data Visualization

One of the final hurdles most researchers encounter is identifying the resources to facilitate the visualization of RNA-seq data. In general, the HTS data visualization tools are not as numerous or as refined as the data analysis tools. Typically, groups either cobble together tools that were likely not intended nor designed to accommodate the visualization of the quantity of data generated by a single HTS data set, much less from multiple data sets, or acquire the expertise to develop their own tools. Our group straddles the two, depending on the project. For larger projects that may have many samples and several collaborators at different locations, we deploy a custom solution that uses a Web-enabled open source genome browser, GBrowse (15), and a custom plugin designed in-house to visualize the data (contact the authors for availability). For smaller projects that may have only a few samples and few or no collaborators, we normally deploy a desktop solution like Tablet (16). Other solutions exist; for example there are other open source programs like AnnoJ (http://www.annoj.org/) and several R-based packages to facilitate RNA-seq data analysis and visualization (http://bioconductor.org/packages/2.7/RNAseq.html). Additionally, there are several commercial HTS software solutions that include the ability to analyze and visualize RNA-seq data; for example, NextGene (http://www.softgenetics.com/NextGENe.html) and Illumina's GenomeStudio (http://www.illumina.com/software/genomestudio_software.ilmn).

4. Notes

1. The computational hardware needed to support a typical RNA-seq project is significant, but not prohibitive. Adequate disk space is most important and is discussed further in the Subheading 3.1. Installed Random Access Memory, or RAM, needed to allow the various programs to run efficiently is moderate, but not unusually large. We recommend a minimum of 5 GB of RAM for smaller projects (for example, less than 50 million reads aligned to a bacterial or small eukaryotic genome), whereas 24–32 GB of RAM should accommodate most larger projects (for example, less than 500 million reads aligned to a large eukaryotic genome). The size and complexity of the reference genome sequence can affect the RAM used by some of the programs. We recommend to always have a "top" window, or equivalent, open to monitor processor and RAM usage. Note that these recommendations are specific to RNA-seq when a reference genome is available. If a project requires de novo transcript assembly, significantly more RAM will likely be necessary and may be prohibitive. It is not unusual for de novo assembly to require hundreds of gigabytes of RAM, which may necessitate specialized computational hardware.

2. In our experience, predicting the amount of disk space required for a particular experiment is notoriously unreliable. New sequencing technologies, kit formulations or vendor-supported throughput recommendations can directly and unexpectedly affect disk space usage. Additionally, different analysis methodologies may appear in the literature that motivate unexpected testing and validation, each of which can affect disk space usage. While working on a project, a constant homeostatic balance of the considerations pertinent to the analysis, backup, and archiving of the data dictates the disk space requirements

and the activities required to most effectively use the available hard drive space. For example, once base calls are generated, the raw data from the sequencer become less important to commit to high-performance disks and may be moved to less expensive archival storage (see discussion in text). If the primary sequencer data is moved to another file system, the gained space can facilitate more analyses. These manifestations of the homeostasis require constant attention and planning to most effectively and efficiently use hard drive space when it is a limiting factor. We recommend planning for 500 GB–1 TB of disk space per sample, depending on the sequencing technology. Illumina trends towards the high end of the scale, whereas Roch/454 trends towards the lower end.

3. Readers are encouraged to consult the online bowtie manual (http://bowtie-bio.sourceforge.net/manual.shtml) for usage directions, with particular attention to bowtie-build and locating the bowtie indexes. Typically, we place all bowtie indexes in a single directory and indicate this directory to the bowtie program by setting the environment variable BOWTIE_INDEXES to that path.

4. Note that cufflinks accepts a single input file. It is tempting to provide the paths to several BAM/SAM files as input arguments, but cufflinks only reads the first file in the list. Thus, it is usually necessary to use samtools to merge multiple datasets. For example, to merge two data sets, data1.bam and data2.bam, invoke samtools as: samtools merge output.bam data1.bam data2.bam.

Acknowledgments

This work was supported by NSF grant 0701731 and a Missouri Life Sciences Trust Fund Research Grant.

References

1. Hannon (2011) FASTX-Toolkit, FASTQ/A short-reads pre-processing tools. http://hannonlab.cshl.edu/fastx_toolkit/index.html. Accessed 25 Feb 2011
2. Li H et al (2009) The sequence alignment/map format and SAMtools. Bioinformatics 25:2078–2079
3. Langmead B et al (2009) Ultrafast and memory-efficient alignment of short DNA sequences to the human genome. Genome Biol 10:R25
4. Trapnell C, Pachter L, Salzberg SL (2009) TopHat: discovering splice junctions with RNA-Seq. Bioinformatics 25:1105–1111
5. Trapnell C et al (2010) Transcript assembly and quantification by RNA-Seq reveals unannotated transcripts and isoform switching during cell differentiation. Nat Biotechnol 28:511–515
6. Langille MG, Eisen JA (2010) BioTorrents: a file sharing service for scientific data. PLoS One. doi:10.1371/journal.pone.0010071

7. Barrett T et al (2011) NCBI GEO: archive for functional genomics data sets–10 years on. Nucleic Acids Res 39:D1005–D1010
8. Parkinson H et al (2011) ArrayExpress update—an archive of microarray and high-throughput sequencing-based functional genomics experiments. Nucleic Acids Res 39: D1002–D1004
9. Ewing B, Green P (1998) Base-calling of automated sequencer traces using phred. II. Error probabilities. Genome Res 8:186–194
10. Longo MS, O'Neill MJ, O'Neill RJ (2011) Abundant human DNA contamination identified in non-primate genome databases. PLoS One. doi:10.1371/journal.pone.0016410
11. Tarailo-Graovac M, Chen N (2009) Using RepeatMasker to identify repetitive elements in genomic sequences. In: Baxevanis AD (ed) Current protocols in bioinformatics, vol Suppl 25. Wiley, New York
12. Schnable PS et al (2009) The B73 maize genome: complexity, diversity, and dynamics. Science 326:1112–1115
13. Richard GF, Kerrest A, Dujon B (2008) Comparative genomics and molecular dynamics of DNA repeats in eukaryotes. Microbiol Mol Biol Rev 72:686–727
14. Vicient CM (2010) Transcriptional activity of transposable elements in maize. BMC Genomics. doi:doi:10.1186/1471-2164-11-601
15. Stein LD et al (2002) The generic genome browser: a building block for a model organism system database. Genome Res 12:1599–1610
16. Milne I et al (2010) Tablet—next generation sequence assembly visualization. Bioinformatics 26:401–402

Chapter 17

Identification of MicroRNAs and Natural Antisense Transcript-Originated Endogenous siRNAs from Small-RNA Deep Sequencing Data

Weixiong Zhang, Xuefeng Zhou, Jing Xia, and Xiang Zhou

Abstract

Next-generation sequencing (NGS) is becoming a routine experimental technology. It has been a great success in recent years to profile small-RNA species using NGS. Indeed, a large quantity of small-RNA profiling data has been generated from NGS, and computational methods have been developed to process and analyze NGS data for the purpose of identification of novel and expressed small noncoding RNAs and analysis of their roles in nearly all biological processes and pathways in eukaryotes. We discuss here the computational procedures and major steps for identification of microRNAs and natural antisense transcript-originated small interfering RNAs from NGS small-RNA profiling data.

Key words: MicroRNA, nat-siRNA, Next-generation deep sequencing

1. Introduction

Many small noncoding RNAs (sncRNAs) have been identified in recent years as essential gene and genome regulators in plants and animals (1–6). They are processed by RNaseIII-type ribonuclease Dicer or Dicer-like (DCL) proteins and incorporated into Argonaute (AGO) proteins in RNA-induced silencing complex (RISC) for their action. The two most studied sncRNAs are microRNAs (miRNAs) and small interfering RNAs (siRNAs). miRNAs are key gene expression repressors functioning mainly at the posttranscriptional level through mRNA cleavage, translation attenuation, or mRNA degradation (5, 6), while siRNAs can act at both the transcriptional level through DNA methylation and histone modification and posttranscriptional level by regulating gene expression (1, 2). In addition to the differences in the small RNA biogenesis pathways for their generation and gene regulatory functions, miRNAs and siRNAs have distinctive structural characteristics. miRNAs are typically processed from RNA polymerase II transcripts that fold back into hairpin structures, while siRNAs are normally produced from double-stranded RNAs. One important group of endogenous siRNAs is the class of siRNAs that are generated from natural antisense transcripts (nat-siRNAs) which have been recognized to play important roles in stress response in plants (7, 8) and gene regulation in animal development (9, 10).

Thanks to the dramatic advance in the next-generation sequencing (NGS) technologies, it is now an effective approach to identify sncRNAs using NGS profiling of small RNA species, followed by computational or bioinformatic analysis of the NGS data. Here, we discuss the key steps and basic procedures for identifying novel miRNAs and nat-siRNAs and quantifying their expression across different samples by analyzing small-RNA data from NGS profiling experiments. The methods discussed here are based on our previous work in this domain, including that presented in (8, 11–17).

2. Data and Programs Used

Several genomic resources and software tools are used for identification and analysis of miRNAs and nat-siRNAs. The genomic resources required include the genome sequences and genome annotation, preferably including ribosomal RNAs (rRNAs), transfer RNAs (tRNAs), small nucleolar RNAs (snoRNAs), small nuclear snRNAs (snRNAs), and miRNAs as well as transposons and repetitive elements. Some existing software tools will be used for mapping sequences, including BLAST or more recent tools such as Bowtie (18), and for analyzing RNA secondary structures, such as RNAfold (19).

3. Methods

3.1. Data Cleaning and Initial Processing

Despite its high quality in data generation, NGS is still error prone, and raw sequence reads from NGS need to be processed to remove erroneous reads, i.e., those that carry no 3′ sequencing adaptors or have low quality. Sequence reads that are shorter than 17 nt are also discarded. The adaptor-trimmed, high-quality reads are referred to as *qualified reads* in the rest of the discussion.

The qualified reads can be classified into different categories depending on the genomic loci from which they may originate by mapping the reads to the genome. In particular, they can be first classified into regions relative to protein-coding genes, i.e., intergenic, intronic, exonic, and 3′ and 5′ untranslated regions. For identification of miRNAs, reads mapped to intergenic regions can be further classified into different categories of various known noncoding genes and transcript units, including rRNAs, tRNAs, snoRNAs, snRNAs, miRNAs, transposons, and repetitive elements. Often, regions encoding these known noncoding genes or transcripts are also discarded. Note that miRNAs may also be processed from snoRNAs, as reported recently (20).

3.2. Identification of Novel miRNAs

The genomic loci that one or more high-quality reads can map to with no more than k mismatches are first collected, forming the initial set of candidate miRNA loci, where k is an integer, often no greater than 2. The rule of thumb for choosing the value of k is that the closer the profiled samples are to the reference genome the smaller the value. In order to focus on novel miRNAs, the loci of known miRNAs are removed. The candidate loci that some qualified reads can map to are processed to merge neighboring candidate loci if they are adjacent to one another within 30 nt. The folding structures around the (merged) loci are then examined to detect candidate miRNA precursors. Since miRNA precursors in animal are typically 80-nt long, 100 nt is thus used as the length of putative pre-miRNAs; likewise, since plant miRNA precursors are normally 150-nt long, 180 nt is the target length for putative pre-miRNAs. At each genomic locus to be analyzed, a series of sequence segments covering the sequence reads are extracted for secondary structure analysis. The starting sequence segment extends approximately 220 nt upstream of each merged locus, and subsequent segments are extracted by a sliding window of 100 or 180 nt, for finding animal or plant pre-miRNAs, respectively, with an increment of, for example, 10 nt, until the window reaches 220 nt downstream of the merged locus. These segments are folded in silico by an RNA folding program, such as RNAfold (19).

Finally, candidate miRNAs are chosen from the folded structures based on the following criteria: (1) a folding structure forms a hairpin with folding energy no greater than −18 Kcal/mol; (2) a stem of a candidate is at least 18-nt long and the number of the highest read on such a stem is greater than a predetermined parameter, e.g., 10; and (3) the presence of at least one raw sequencing read corresponding to a possible miRNA* sequence, and the presence of possible 2- or 3-nt 3′ overhangs on the miRNA/miRNA* duplex. The rationale for the third criterion is that miRNA precursors are known to be processed by RNase III enzyme, Dicer, yielding a duplex of ~22-nt miRNA/miRNA* duplexes with 2–3-nt 3′ overhangs. We note that if experimental analysis follows the sequencing profiling and bioinformatic analysis the third criterion may be ignored.

3.3. Analysis of miRNA Expression

3.3.1. Detection of miRNA Expression

Qualified reads are mapped to the genomic loci of annotated miRNAs with perfect matches or no more than two mismatches. The total number of the mapped reads that start within the interval of six nucleotides centered around the annotated starting genomic locus of a miRNA sequence is then taken as the raw expression level of the miRNA. The purpose of using an interval for a starting genomic locus is to accommodate isoforms of a miRNA (17, 21) due to imprecise Dicer activities (22). Reads that can map to multiple miRNA loci are attributed to all potential derivative miRNAs.

3.3.2. Analysis of Digital miRNA Expression Abundance

The miRNA raw qualified read counts are then normalized to adjust for variation in the overall profiling depth of a small RNA library. Let m be the total number of qualified reads mapped to the genome and w the raw read count of a miRNA. The digital expression level, or normalized expression level, of the miRNA is then $w \times n/m$, where n is a large constant, e.g., one million, set for all small-RNA libraries for the purpose of comparing relative abundances of the miRNA across multiple libraries. The value for n may also be the average of the qualified reads in all small-RNA libraries that are mapped to the genome.

3.3.3. Identification of Differentially Expressed miRNAs

In a disease study or a study of stress response in plants, it is of primary importance to identify miRNAs that are differentially expressed across two sets of samples, i.e., patients vs. normal controls or stressed vs. untreated plants. To this end, many statistical methods developed previously for identification of differentially expressed genes from microarray data, such as SAM (23) and Rank Product (24), can be adopted by using digital expression abundance.

3.4. Identification of Endogenous siRNAs Originated from Natural Antisense Transcripts

3.4.1. Identification of Cis- and Trans-Natural Antisense Transcripts

The key to nat-siRNA finding is the identification of natural antisense transcripts. If two transcripts reside on the opposite strands at the same genomic locus and the overlapped region is longer than n-nt, they have the potential to form a pair of *cis*-natural antisense transcripts or *cis*-NATs, where n is a parameter typically with a value no less than 23 to support the generation of at least one siRNA sequence. A more reasonable choice for n is three or four times the length of siRNAs. Note that the transcripts considered may be protein-coding genes, pseudo genes, or nonprotein-coding transcripts.

According to the directions of the involved transcription units, *cis*-NATs can be categorized into three groups, *convergent cis*-NATs with 3′ ends overlapping, *divergent cis*-NATs with 5′ ends overlapping, and *enclosed cis*-NATs with one transcription unit being entirely overlapped by the other.

Trans-natural antisense transcripts, or *trans*-NATs, are transcripts that originate anywhere in the whole genome and parts of the transcripts can form perfect complementary pairings. They can be identified by searching for pairs of transcription units that share regions of sequence that are perfect complements to each other. Specifically, two transcripts from two different genomic loci are considered to have the potential to form a *trans*-NAT pair if they satisfy the following two criteria: they have a continuous perfect complementary pairing region longer than m-nt and their overlapping region can form an RNA–RNA duplex. The parameter m should be sufficiently large, e.g., 100, to support possible RNA–RNA annealing. This possibility can be examined in silico by the computational tool, DINAMelt (25, 26), which inspects whether the overlapping regions of a *trans*-NAT pair can melt into an RNA–RNA duplex.

3.4.2. nat-siRNAs and Their Enrichment

With NATs (*cis*- or *trans*-NATs) given, putative nat-siRNAs are represented by the qualified reads that map perfectly to the overlapping regions of the NATs.

The enrichment of the reads in the overlapping regions of NATs can be quantified by their densities and statistical significance. For each pair of NATs, denote N_o and N_g to be the numbers of qualified reads mapped to the overlapping region and the nonoverlapping regions of the two transcripts, respectively, and let L_o and L_g be the length of the overlapping region and the total length of nonoverlapping regions of the two transcripts, respectively. The density of qualified reads in the overlapping region is then N_o/L_o, and that in the nonoverlapping regions is N_g/L_g. The average densities in the overlapping regions (A_o) and the nonoverlapping regions (A_g) are computed. The ratio A_o/A_g is then considered as the degree of the enrichment. The enrichment of sequencing reads in all overlapping regions, relative to all nonoverlapping regions, can be tested using the one-tail paired two-sample t-test using the two sets of paired samples of small RNA density, N_o/L_o and N_g/L_g. The p-value of the t-test is then the statistical significance of the enrichment.

In order to gain further statistical significance of the small RNAs in the overlapping regions of NATs, the same method can be extended to compare the enrichment of siRNAs in the overlapping regions to that in other regions of the genome, e.g., a set of arbitrarily chosen regions of protein-coding but non-NAT transcripts.

4. Note

We like to mention that many parameters mentioned above, such as minimal fold energy and sliding window size for novel miRNA finding, are empirical and may have to be considered case by case depending on the organisms and particular applications under consideration.

Acknowledgment

This work was supported in part by NSF grant DBI-0743797, NIH grants RC1AR058681 and R01GM086412, as well as an internal grant of Fudan University to W. Zhang.

References

1. Baulcombe D (2004) RNA silencing in plants. Nature 431:356–363
2. Ghildiyal M, Zamore PD (2009) Small silencing RNAs: an expanding universe. Nat Rev Genet 10:94–108
3. Malone CD, Hannon GJ (2009) Small RNAs as guardians of the genome. Cell 136:656–668
4. Moazed D (2009) Small RNAs in transcriptional gene silencing and genome defence. Nature 457:413–420
5. Bartel DP (2004) MicroRNAs: genomics, biogenesis, mechanism, and function. Cell 116: 281–297
6. Kim VN, Nam J-W (2006) Genomics of microRNA. Trends Genet 22:165–173
7. Borsani O, Zhu J, Verslues PE, Sunkar R, Zhu J-K (2005) Endogenous siRNAs derived from a pair of natural cis-antisense transcripts regulate salt tolerance in Arabidopsis. Cell 123: 1279–1291
8. Zhou X, Sunkar R, Jin H, Zhu J-K, Zhang W (2009) Genome-wide identification and analysis of small RNAs originated from natural antisense transcripts in Oryza sativa. Genome Res 19:70–78
9. Okamura K, Lai EC (2008) Endogenous small interfering RNAs in animals. Nat Rev Mol Cell Biol 9:673–678
10. Watanabe T, Totoki Y, Toyoda A, Kaneda M, Kuramochi-Miyagawa S, Obata Y, Chiba H, Kohara Y, Kono T, Nakano T, Surani MA, Sakaki Y, Sasaki H (2008) Endogenous siRNAs from naturally formed dsRNAs regulate transcripts in mouse oocytes. Nature 453: 539–543
11. Sunkar R, Zhou X, Zheng Y, Zhang W, Zhu J-K (2008) Identification of novel and candidate miRNAs in rice by high throughput sequencing. BMC Plant Biol 8:25
12. Jagadeeswaran G, Zheng Y, Li Y-F, Shukla LI, Matts J, Hoyt P, Macmil SL, Wiley GB, Roe BA, Zhang W, Sunkar R (2009) Cloning and characterization of small RNAs from Medicago truncatula reveals four novel legume-specific microRNA families. New Phytol 184:85–98
13. Jagadeeswaran G, Zheng Y, Sumathipala N, Jiang H, Arrese EL, Soulages JL, Zhang W, Sunkar R (2010) Deep sequencing of small RNA libraries reveals dynamic regulation of conserved and novel microRNAs and microRNA-stars during silkworm development. BMC Genomics 11:52
14. Reddy AM, Zheng Y, Jagadeeswaran G, Macmil SL, Graham WB, Roe BA, Desilva U, Zhang W, Sunkar R (2009) Cloning, characterization and expression analysis of porcine microRNAs. BMC Genomics 10:65
15. Zhang W, Gao S, Zhou X, Xia J, Chellappan P, Zhou X, Zhang X, Jin H (2010) Multiple distinct small RNAs originate from the same microRNA precursors. Genome Biol 11:R81
16. Chellappan P, Xia J, Zhou X, Gao S, Zhang X, Coutino G, Vazquez F, Zhang W, Jin H (2010) siRNAs from miRNA sites mediate DNA methylation of target genes. Nucleic Acids Res 38:6883–6894
17. Reese TA, Xia J, Johnson LS, Zhou X, Zhang W, Virgin HW (2010) Identification of novel microRNA-like molecules generated from Herpesvirus and host tRNA transcripts. J Virol 84:10344–10353
18. Langmead B, Trapnell C, Pop M, Salzberg SL (2009) Ultrafast and memory-efficient alignment of short DNA sequences to the human genome. Genome Biol 10:R25
19. Hofacker I, Fontana W, Stadler P, Bonhoeffer S, Tacker M, Schuster P (1994) Fast folding and comparison of RNA secondary structures. Monatshefte f Chemie 125:167–188
20. Brameier M, Herwig A, Reinhardt R, Walter L, Gruber J (2010) Human box C/D snoRNAs with miRNA like functions: expanding the range of regulatory RNAs. Nucleic Acids Res 39:675–686
21. Morin RD, O'Connor MD, Griffith M, Kuchenbauer F, Delaney A, Prabhu A-L, Zhao Y, McDonald H, Zeng T, Hirst M, Eaves CJ, Marra MA (2008) Application of massively

parallel sequencing to microRNA profiling and discovery in human embryonic stem cells. Genome Res 18:610–621

22. MacRae IJ, Doudna JA (2007) Ribonuclease revisited: structural insights into ribonuclease III family enzymes. Curr Opin Struct Biol 17:138–145

23. Tusher VG, Tibshirani R, Chu G (2001) Significance analysis of microarrays applied to the ionizing radiation response. Proc Natl Acad Sci USA 98:5116–5121

24. Breitling R, Armengaud P, Amtmann A, Herzyk P (2004) Rank products: a simple, yet powerful, new method to detect differentially regulated genes in replicated microarray experiments. FEBS Lett 573:83–92

25. Markham NR, Zuker M (2005) DINAMelt web server for nucleic acid melting prediction. Nucleic Acids Res 33:W577–W581

26. Dimitrov RA, Zuker M (2004) Prediction of hybridization and melting for double-stranded nucleic acids. Biophys J 87:215–226

INDEX

A

Adapter 2–8, 10–12, 15, 48, 54, 68, 72, 158, 171, 172, 183, 184, 186, 187, 190, 212, 213
AGO. *See* Argonaute
Alignment 40, 101–104, 183, 187, 188, 194, 196, 197, 205, 206, 214, 215
Amplification 11, 16, 68, 71, 73, 92, 99, 100, 108, 173, 183–185, 197
Analysis
    computational 201–218
    gene expression 1–17
Annotation 1, 2, 14, 98, 106, 108, 193, 195, 197, 213, 222
Aphid 47–50
Arabidopsis 75–85, 87, 88, 90–94, 143, 151, 165, 166, 169, 178
Argonaute (AGO) 165, 178, 221
Assembly, sequence 206

B

Bioinformatics 180, 187–188

C

cDNA 1–4, 6–9, 14–16, 22, 27, 35, 48, 53, 54, 61, 63–66, 68, 71, 73, 87, 99, 100, 133, 156, 173, 184–187, 193, 194
Contig 188, 195

D

Data processing 98, 107, 190, 205, 206, 213, 215, 222–223
Detection, colorimetric 76
Dicer and Dicer-like 165, 221
Digoxigenin 21, 76, 146

E

Electrophoresis, polyacrylamide gel 11, 26–29, 32, 115–116, 122
EST. *See* Expressed sequence tag
Exon 101, 197
Expressed sequence tag (EST) 1, 194–197

F

Filamentous fungi 156
Fluorophore 112
Fusarium oxysporum 156

G

Gall 88, 90, 94
GenBank 14, 189
Gene expression 1–17, 20, 37, 59, 75, 76, 98, 107, 155, 165, 178, 204, 221
Genome 1, 2, 13, 34, 47, 48, 59, 60, 66, 71, 72, 75, 97–99, 101, 104–107, 123, 131, 132, 137, 178, 181, 187–190, 197, 201, 202, 206, 207, 210, 212–214, 216, 217, 221–225

H

HTS. *See* Sequencing, high-throughput
Hybridization, in situ 75–85, 144, 145, 151, 152

I

Immunoprecipitation 166, 169–171
Innate immunity 166
Intron 101–103, 123, 212, 215

L

Laser 87–95
Library 47, 48, 55, 56, 60, 61, 66–73, 87, 98–100, 108, 155–164, 167–168, 171–175, 180, 183, 184, 187, 193, 194, 201, 209, 212, 215, 216, 224
LNA. *See* Locked nucleic acid
Locked nucleic acid (LNA) 21, 24, 32, 143–152

M

Malaria 59–73
Mercury 112, 113, 115–119
Mercury, organo- 111–119
Microarray 22, 59, 87, 88, 201, 224
Microtome 77, 80, 146, 149, 152
miRNA. *See* RNA, micro-

mRNA
  isoform 106, 107
  spatial expression 144
Multiplexing 2, 6, 13, 16, 66, 108

N

National Center for Biotechnology Information
  (NCBI) 204
Natural antisense transcripts 222, 224–225
NCBI. *See* National Center for Biotechnology Information
Nematodes 88
Neurospora crassa 156
NGS. *See* Sequencing, next-generation
Nomenclature, file 207, 209
Northern blot 1, 19–43, 122, 132, 138, 139, 144, 174

O

Open reading frame 122, 123

P

PCR
  real-time 1, 23, 122, 132
  reverse transcription 122, 132, 162, 183, 184
Phosphorothioate 111, 112
Pipeline 107, 188,
  194, 196, 197, 202, 203, 207–209, 211, 213, 214
Polyadenylation, alternative 122, 123
Polymorphism 194
Pre-mRNA 97, 122
Profiling, transcript 1, 19, 37, 194
Program 13, 53, 64, 67–69,
  105, 139, 174, 178, 180, 187, 188, 196, 202, 204,
  206, 207, 209, 211–213, 215, 217, 218, 222, 223
Promoter 19, 34, 80, 84,
  121, 123, 135, 137, 140, 144
Pythium ultimum 194

Q

Quality control 55, 170, 175, 187,
  197–198, 205–206, 210

R

Reads
  paired-end 105, 207, 214
  sequence 7, 13, 14, 196, 205–206, 211, 222, 223
Reverse transcriptase 3, 8, 35,
  49, 53, 55, 71, 99, 168, 180, 185, 197
Ribo-depletion 98–99
Ribonuclease 71, 76, 121, 124, 133, 221
Ribonuclease protection assay (RPA) 121–123
Riboprobe 34, 133–134, 138
Ribosomal RNA (rRNA) 37–41, 43, 98, 190, 206
Ribozyme 111, 112

RISC. *See* RNA-induced silencing complex
RNA, catalytic 111
RNA, cytoplasmic 124–127, 178
RNA extraction 39, 40, 49–50,
  55, 61–62, 88, 89, 92, 94, 95, 132, 134–135, 157,
  159, 167, 170–171
RNAfold 222, 223
RNAi. *See* RNA interference
RNA-induced silencing complex
  (RISC) 165, 166, 221
RNA interference (RNAi) 112, 177, 178
RNA, micro- 32, 75, 144,
  166, 178, 183, 187, 190, 223–225
RNA, polyA 62–63, 98, 108
RNA polymerase 80, 81,
  124, 127, 133–135, 137, 140, 222
RNA probe 20, 21, 34,
  36, 37, 121, 122, 127, 129, 133–134, 137–138, 144
RNA processing 123, 183
RNA, satellite 131–140
RNA-seq 47–56,
  59–73, 97–109, 196, 201–218
RNA, small 20–27, 32–34, 36–38, 155–164,
  166–168, 171–175, 180, 182–186, 188, 221–225
RNA, small interferings 155, 165, 178, 221
RNA, sulfur-containing 111–119
RNA, total 8, 20–32, 35, 37, 40,
  49–52, 56, 61–63, 70, 98, 124–126, 129, 134–137,
  139, 140, 157, 159–160, 163, 181, 184, 190
RNA, viral 131–140
RPA. *See* Ribonuclease protection assay
rRNA. *See* Ribosomal RNA

S

SAGE. *See* Serial analysis of gene expression
Satellite RNA. *See* RNA, satellite
Script 13, 102, 105, 188, 197, 205, 211–214
Section, cryo- 87–95
Section, tissue 75, 76, 80, 87
Sequencing
  AppliedBiosystems or SOLiD 2, 7, 8,
    10–12, 14, 17
  deep 59, 155–164,
    166, 175, 177–190, 193, 194, 221–225
  high-throughput 21–23, 59,
    97, 155–164, 179, 205
  Illumina or Solexa 47–56,
    100, 156, 180, 187, 196
  massively parallel 2, 6, 97
  next-generation 21, 47, 97, 178, 193, 222
  pyro-or 454 or Roche 47, 100, 193, 194
  Sanger 47, 55, 193, 194
  short-read 2, 105, 204
Serial analysis of gene expression (SAGE) 1–17

Single nucleotide polymorphism (SNP) 194
siRNA. *See* RNA, small interferings
SNP. *See* Single nucleotide polymorphism
Software 13, 42, 116, 180, 181, 187, 190, 202, 204, 205, 211, 217, 222
Splicing, alternative 97–109, 122, 123, 178
sRNA. *See* RNA, small

T

Tissue
    embedding 76, 79, 89, 90, 148
    fixation 76, 78, 79, 88–90
Tomato 88, 92, 93, 95, 156, 166
Transcript, abundance of 22, 37, 105

Transcription, in vitro 25, 30, 34, 77, 80–81, 124–127, 129, 132–133, 135, 140
Transcriptome 1, 7, 98, 101, 105, 107, 108, 178, 194
tRNA 37, 77, 78, 81, 111, 112, 125, 127, 128, 133, 138, 147

U

Uridine 111

V

Viral RNA. *See* RNA, viral
Virus
    Grapevine vein clearing 179
    parvo 122, 123

Printed by Printforce, the Netherlands